2009 China Interior Design Annual

2009中国室内设计年鉴（2）

《中国室内设计年鉴》编委会

辽宁科学技术出版社

目录

餐厅 Restaurant

样板 Sample House

商业展示 Exhibition

向阳渔港·文鼎一号 Wending No.1 of Sunward Fishery Restaurant

设计单位:宁波汉文设计 / 设计:万宏伟 / 面积:2500 m² / 主要材料:石材、柚木饰面、钢板刻画、手绘刺绣、手绘壁、香槟金镜 / 坐落地点:宁波 / 工程造价:1000万元 / 完工时间:2009年7月 / 摄影:刘鹰

Design Unit:Han Wen Design / Design:Wan Hongwei / Area:2500 m² / Main Materials:stone、teak veneer、steel plate、embroidery、gold mirror/ Location:Ningbo / Project Cost:RMB 10 000 000 / Finish Time:Jul,2009 / Photography:Liu ying

文鼎一号是向阳渔港打造的高端餐饮品牌，分大堂公共就餐区和风格各异的包间。公共就餐区注重空间动线流向的指引和划分，偌大的灯罩围合出空间内空间的感觉。户型的平面布局让玻璃发挥了巨大的作用，不论是包间还是大堂，都能饱览室外美景。手绘刺绣的见缝插针为空间平添几分秀色。二层超豪华包间更因层高优势可以欣赏到一线江景。

Wending No. 1 is a top-level brand restaurant created by Sunward Fishery Restaurant, including the public dining area in the hall and compartments with different styles. Space guidance is emphasized in the public dining area, and large lampshades create new spaces in the existed ones. Thanks to the horizontal layout, glass makes it easier for customers to enjoy the beautiful scenery outside both in the hall and compartments. Manual embroiders everywhere make the space even more enjoyable. You can enjoy the view of the river in the super-luxurious compartments on the second floor.

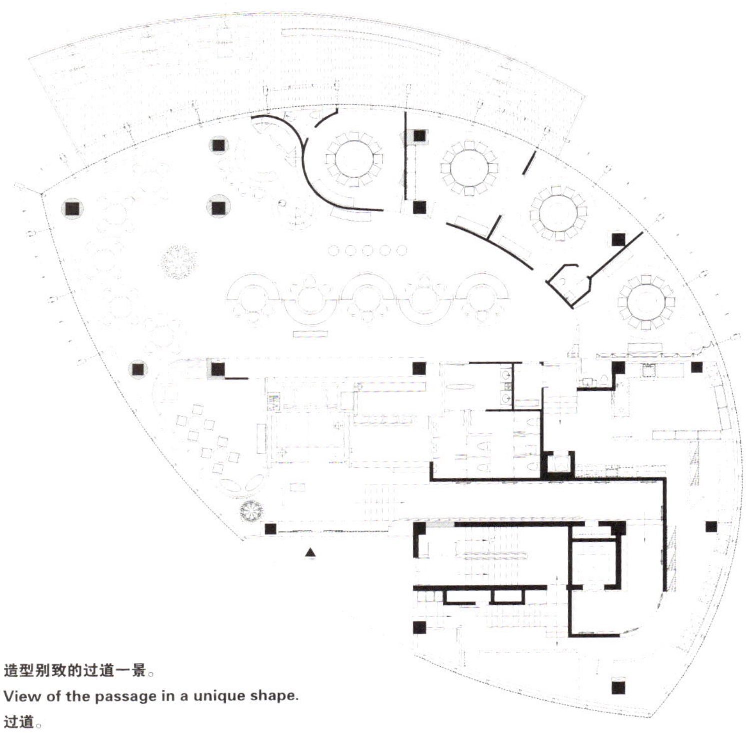

造型别致的过道一景。
View of the passage in a unique shape.
过道。
Passage.

CONTENTS

偌大的灯罩围合出小小世界。
A small world in the large lamp screen.

华贵雍容的水晶灯。
Luxurious and graceful crystal lamp.
经典的走廊。
Classic corridor.

风格各异的包间设计。
Compartment design in various styles.
纯白的空间配上缤纷的花纹，自然流畅。
Pure white space with colorful patterns, natural and fluent.

二层豪华包间可以饱览一线江景。
The front line river landscape can be viewed from the top grade compartment in the second floor.
大堂就餐区落地玻璃坐收室外美景。
The outdoor landscape can be viewed via the French windows in the dining zone of the hall.

会宾楼 Huibin Building

设计:费宁 / 面积:6000 m^2 / 主要材料:大理石、镜面、仿金属玻璃、皮革、金影木 / 坐落地点:合肥 / 工程造价:9000万元 / 完工时间:2009年4月 / 摄影:文宗博

Design:Fei ning / Area:6000 m^2 / Main Materials:marble、mirror surface、metal-like glass、leather、figured / Location:Hefei / Project Cost:RMB 90 000 000 / Finish Time:Apr,2009 / Photography:Wen Zongbo

现代奢华的设计语言，极具震撼的设计效果，夸张尺度的设计符号，各种高档的材质物料，这一切尽在这个项目中体现。灯光在这里烘托了环境气氛，灯具除了照明还有了很强的造型效果。进口石材的运用，虽然增加了成本，但显现出的是花纹图案般的效果。每件饰品都设计了展示效果，精致奢华。晶莹剔透的红酒架，如珠宝盒般夺目。

It has contemporary luxury design language, shocking design effect, design characters with exaggerated dimension and various kinds of high grade materials. All these are reflected in this item. Here the lighting sets off the environmental atmosphere. Apart from lighting functions, the lamps also have strong molding effect. Although the cost increases due to the use of imported stones, but it can show the lathework effect. Every ornament has designed the exhibition effect and it is very delicate and luxury. The glittering and translucent carving red wine bracket dazzles the eyes like a jewel case.

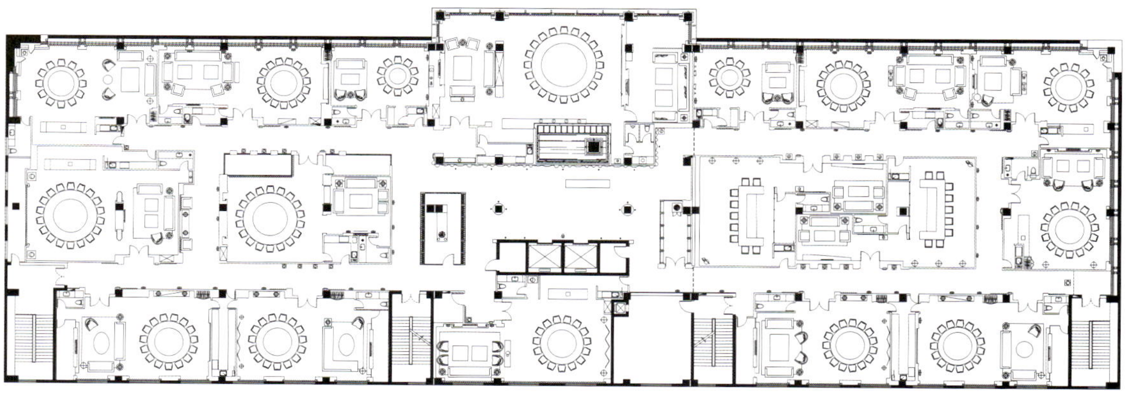

极具震撼的设计效果。
The extremely shocking design effect.
上下贯通的空间动线。
The dynamic lines in the space connecting up and down.

夸张尺度的设计符号。

The dimension-exaggerated design symbols.

现代奢华的设计语言。

The modern luxurious design languages.

节节高升的圆形吊灯。

The successively ascending round pendant lamps.

通长的落地玻璃削弱了室内外的界限。

The thoroughly long floor glass has weakened boundary between the indoor and outdoor.

精致闪亮的吊灯整齐划一。
The delicate and shining pendant lamps are arranged well-ordered in line.
灯光在这里烘托了环境气氛。
The light here has set off the ambient atmosphere.
符号化的设计贯穿始终。
The symbolized design runs through all along.

阿林鼎满香餐厅 A Lin Ding Man Xiang

设计单位:深圳市汇博环境设计有限公司 / 设计:曹成 / 面积:1520 m² / 主要材料:不锈钢水波纹板、水晶砖、艺术玻璃、自流平地面、亚克力制品 / 坐落地点:西安 / 工程造价:800万元 / 完工时间:2008年11月 / 摄影:WISON(泰国)

Design Unit:Shenzhen hope-box / Design:Cao Cheng / Area:1520 m² / Main Materials:stainless steel、crystal brick、fancy glass、self-leveling floor、acrylic products / Location:Xi'an / Project Cost:RMB 8 000 000 / Finish Time:Nov, 2008 / Photography:WISON (Thailand)

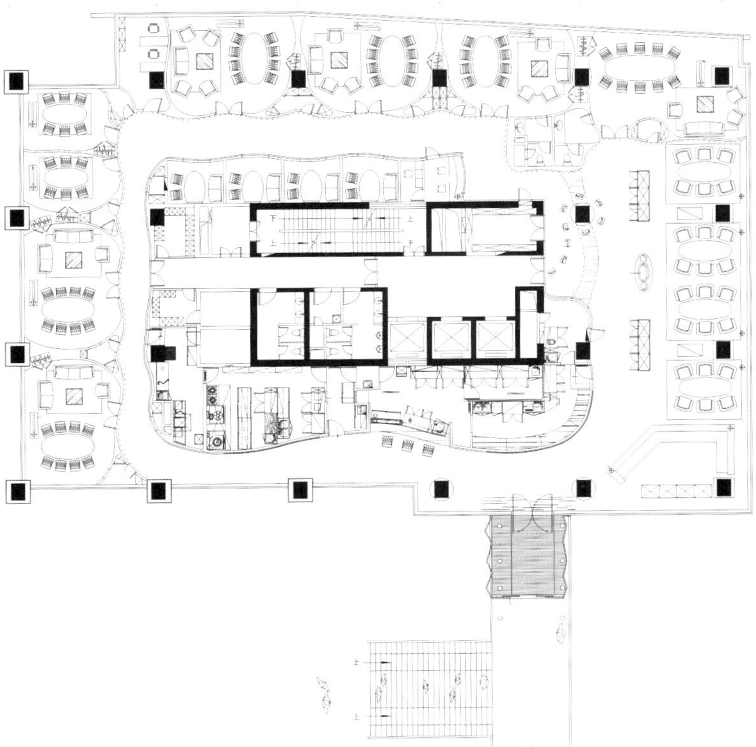

走廊的弧线。
The arc of the corridor.

本案位于西安南门广场旁的CBD玻璃建筑群内，直接面对着久负盛名的、迄今保存最为完整的南门古城墙，基地的现代风格与古老的城墙形成了鲜明的对比。

餐厅主入口为新加建的钢结构廊桥，门廊雨篷由冰凌状的水晶玻璃构成，白色钢构件在超白玻璃间撑起不规则的三角形。由于门廊凹入隐蔽在两栋办公塔楼之间，很不起眼，所以利用形态各异的一群流体不锈钢企鹅雕塑，一溜排开成进出餐厅不同形态，将开始有着些许好奇感的客人引上廊桥。步入廊桥雨篷，一眼就看到了雕刻着两片荷花的水晶玻璃墙面，白色，剔透，没有多余的杂色，这道流动的水晶弧形墙将最难处理的厨房部分完全隐藏起来。

大厅除了吧台外只设置了四围椭圆形的餐台，由于面临城墙的良好观景视角，所以经营方准备在晚间9：40后调暗灯光，更换场景，将用餐转变为酒吧功能。酒吧台和红酒柜依然沿用了水晶玻璃，就连入口处的两根柱子，也被特制的亚克力冰块环抱，浸透着冰色的白光。需要偏重提及的是天花板的处理，为了减弱低标高天花的压抑感，结合主题立意，利用镜面不锈钢开模特制了金属水纹板，很好地解决了标高限制这一硬伤，空间得以反射拉高。同时，为了与其呼应，自流平地面用自创的添加剂调制出水状纹理，此时，立面的水晶玻璃自然成为极具冰感的发光体，在天地间形成的倒影，犹如平静的水面被拨弄开了，其中走动的人影，在这种叠荡中相互融合，尤其从室外看进来，一个流淌的世界得以呈现，此时冰即水，水成冰。

并非一切都是冷色的，作为重点营业区域的包间就用“牡丹”点起了一片片的“绚烂”。8个VIP空间，由不规则的弧线构成，8种色调、8类牡丹、8个典故。包间用唐诗宋词中对“牡丹”的赞誉而命名，牡丹被雕刻在19cm超白弧形水晶玻璃上，利用玻璃前后不同的打光方式使不同形态及地域特点的牡丹得以展现，各色的花蕊绽放着不同的热情迎接着来客。需要说明的是，在雕刻玻璃的制作过程中，我们克服了钢化工艺对不同雕刻深度产生的弧形变形不一致的问题，力争玻璃墙面的弧度过渡自然，使外冷内暖在此空间的过渡一目了然。

水晶玻璃墙面与带状雾化不锈钢中点缀的水晶水珠特有的晶透感在天花和地面的反射下得以夸大，天花上的水浸透入墙面，加上鱼池那两抹蓝色的荡漾，让空间的延伸感在流淌中反复展现。

This project locates at the CBD glass architectural complex by South Gate Square of Xi' an, China, directly fronting the renowned and well-reserved South Gate ancient city wall. Its modern style presents a sharp comparison to the delegate ancient flavor of the wall.

The main entrance of the restaurant is a newly built gallery bridge in steel structure, with above weather shed made of ice slush shaped crystal glass. The glass is supported by sets of white steel members, in shape of irregular triangles. Since the location of the entrance is not by the main road and easy to miss, some vivid steel penguin statues are placed along the staircase below the gallery bridge to lure curious potential customers follow their leads. Once set foot on the gallery bridge, customers are greeted by a white crystal glass wall with two lotuses sculpted on. Their leaves, seedpods and falling petals artistically outline the contrast between peaceful water and dynamic lotus. The flowing arc crystal wall encloses the core tube of the restaurant, perfectly concealing the kitchen at the same time, which otherwise is a difficulty to deal with.

Well located as it is, the project' s internal conditions, however, impose restraints on the designing, as the standard floor height is only 2900mm and its conduit is intricate, especially the complicated pipelines in the kitchen which considerably confines the gross area of it. Relying on mass calculation and well-planned adjustments on vent, ventilation, fire protection and various bridges, we manage to keep floor-to-ceiling height no less than 2780mm with plumbing being totally concealed. These work which are easily neglected by client consumes lots of mental efforts and time.

Apart from a bar counter, there are only oval-shaped tables set in the lobby. In order to take full advantage of the nice view of the ancient wall, the restaurant plans to dim lights after 9:40pm and change settings, easily turning it into a bar. Crystal glass is adopted once again in building up the bar counter and the wine cabinet, and the two pillars at the entrance are encased by custom acrylic ice, sparkling with white light.

What needs to be mentioned in particular is the design of the ceiling. Mirror surfaced stainless steel is adopted to make custom metal ripple board which extends the space with its reflection, therefore minimizing the pressure from the relatively low floor-to-ceiling height. In the meantime, self-made additive is applied to develop concrete flooring veined water grain to go with the ceiling, making the standing crystal wall a natural cool sparkling subject. The inverted images of the crystal wall in the ceiling and the flooring make the space look like wavy water surface, absorbing the images of people walking along consummately. Looked from outside, a flowing world comes into being, a beautiful mixture of ice and water.

However, not everything is in cold color. As key business area, private rooms are decorated by splendid "peonies". 8 VIP rooms are in 8 colors, with 8 kinds of peonies and 8 different literary quotations. Each VIP room is named after genuine admirations to peonies picked from ancient poems. Each peony is engraved upon 19mm white arc crystal glass, exhibiting characteristic of its own variety and region, warmly welcoming all customers. In respect that Tibetan religion holds special interpretation to peony, Tibetan words are rendered in floor blanket of each VIP room, giving further explanation to each room's name. In the process of glass carving, we manage to make the radian of glass as natural as possible, overcoming the problem of uneven arc deformation caused by toughening treatment on carving of different depth, making the interim from cold to warm, outside to inside clear at a glance.

The sparkling sense of the crystal glass wall surface and the crystal water drops on the ribbon shaped stainless steel is intensified by the reflection function of both ceiling and flooring. Together with "water" from the ceiling soaking into crystal wall, and the undulating blue color from the fish jars, the theme of stretched space is enhanced.

室外入口雕塑。
The outdoor sculpture at the entrance.
挑空鱼缸。
The fish jar.
曲线灵动的回廊。
The cloister with dynamic curves.

大厅入口，如进入冰城一般。
The entrance of the lobby seems like entering an ice city.

有区别但又统一风格的VIP包厢。

The VIP rooms that are distinguished from each but have a unified style.

86° 咖啡桌面游戏主题店 86°Coffee Desktop Game Shop

设计单位:南京广暄工程设计顾问有限公司 / 设计:陈广暄、马珺 / 参与设计:曹海军、张乃伟 / 面积:600 m² / 主要材料:铁刀木防潮板、仿皮纹墙地砖、复合木地板、艺术玻璃 / 坐落地点:南京 / 工程造价:75万元 / 完工时间:2008年8月 / 摄影:南京远景摄影设计有限公司

Design Unit:Nanjing Guangxuan Design / Design:Chen Guangxuan、Ma Jun / Participatory Design:Cao Haijun、Zhang Naiwei / Area:600 m² / Main Materials:cassia siamea、floor tile、compound floor board、fancy glass / Location:Nanjing / Project Cost:RMB 750 000 / Finish Time:Aug,2008 / Photography:Nanjing YuanJing photograph Design Co., Ltd.

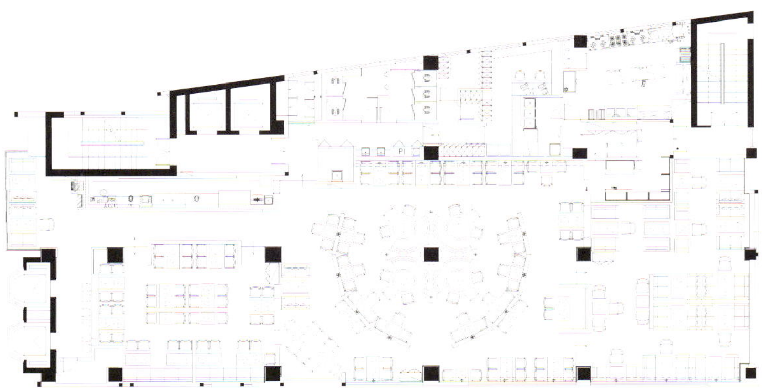

在整个设计中，各种餐具的形象以不同的形式表现出来成为主角，含蓄的餐具（叉子、筷子)隐身在了墙面的条码图案中。大方的器皿出现在吧台侧面的连笔画中，如同一场T台表演秀。墙面上盘子与勺子的组合装饰，像一场小型的交谊舞会。半高隔断里吊挂着整排勺子，在灯光的照射下正在演奏一场动人的打击乐。分布在空间里的场景橱窗中所有的器物以一种超现实的方式站立在墙面上。一本翻开的书、一盘没下完的棋、刚刚倒好的红茶、一场鏖战中的牌局。似乎场景中的主人从未离开。所有这一切对餐具的总动员都让人感觉到四周充满了生命跃动的气息。

In the whole design, the images of different table wares are displayed in different forms and play the leading roles. The connotative table wares (forks and chopsticks) are hidden in the bar code figure on the wall. The decent household utensils appear in the continuous stroked picture on the side of the bar counter, just like a T-stage show. The decorative combination of the dishes and spoons on a wall looks like a small dancing party. The whole row of spoons is hanging inside the half-height partition wall and is playing moving percussion music under the illumination of the light. All the appliances in the scenic shop windows distributed in the space are standing on the wall in a surrealistic way: an opened book, an ongoing chess play, a just filled red tea and a hard-fighting card play, which seem that the hosts have never left. All these on the overall mobilization of the table wares are delivering a lifeful atmosphere.

分布在空间里的场景橱窗中所有的器物以一种超现实的方式站立在墙面上。
All the appliances in the scenic shop windows distributed in the space are standing on the wall in a surrealistic way.

半高隔断里吊挂着整排勺子，在灯光的照射下正在演奏一场动人的打击乐。
The whole row of spoons is hanging inside the half-height partition wall and is playing moving percussion music under the illumination of the light.
墙面上盘子与勺子的组合装饰，像一场小型的交谊舞会。
The decorative combination of the dishes and spoons looks like a small dancing party.
大方的器皿出现在吧台侧面的连笔画中。
The decent household utensils appear in the continuous stroked picture on the side of the bar counter.
各种餐具的形象以不同的形式表现出来成为主角。
The images of different table wares are displayed in different forms and play the leading roles.

滨江花园酒店 Riverside Garden Hotel

设计单位:南京广暄工程设计顾问有限公司 / 设计:陈广暄、马珺 / 面积:5300 m² / 主要材料:双色竹片板、抛光砖、壁纸、灌胶竹节板、米黄大理石 / 坐落地点:南京 / 工程造价:2200万元 / 完工时间:2009年3月 / 摄影:南京远景摄影设计有限公司

Design Unit:Nanjing Guangxuan Design / Design:Chen Guangxuan、Ma Jun / Area:5300 m² / Main Materials:bamboo board、polished tiles、wallpaper、corrugated board、marble / Location:Nanjing / Project Cost:RMB 22 000 000 / Finish Time:Mar,2009 / Photography:Nanjing YuanJing photograph Design Co., Ltd.

该项目为南京知名餐饮企业“天狮百盛”设于绿化博览园内的餐饮分店。园内景观皆出自全国各省园林名家之手，这份先天优势无疑成为设计师最希望保留并予以借用的设计资源。如何以最贴近自然的清新风格塑造成适合于高端消费对象的场所，成为本次设计的主旨。最终设计师在“大气”和“灵气”之间求得了平衡。大气来自对空间尺度的把握，宽阔的平面划分和充足空间高度给人以豪放心理感受。而灵气则表现在对自然植物形态和色彩的再现，以浅灰的色彩基调为铺垫，使得原本质朴无奇的草绿、橙黄、桃红等自然色变得绚丽而时尚。

It is a branch restaurant invested by in the famous catering enterprise in Nanjing – “Lion Parkson” in Hexi Greenery Theme Park. The sights in the garden are done by the famous gardeners from many provinces around the county. This innate advantage doubtlessly becomes the design resource the designer mostly hopes to maintain and borrow for use. How to create a sense of place suitable for the high-end consumption targets by a fresh style closest to the nature has become the main principle of this design. Finally, the designer has reached the balance between “grandness” and “charm”. The grandness comes from the control over the space dimensions. The wide spacious plan layout and the sufficient space height make people bond and unconstrained while the charm is revealed by the reproduction on the forms and colors of the natural plants. The light gray color is taken a background so that the natural colors such as grass green, yellow orange and pink that used to be simple and common become splendid and fashionable.

原本质朴无奇的草绿变得绚丽而时尚。
The grass green that used to be simple and common becomes splendid and fashionable.
橙黄的墙面和吊灯赋予空间十足暖意。
The orange wall and pendant lamp endow the space with full warmness.

不同形式的照明方式丰富了空间的层次。
The different lighting methods have enriched the space levels.
开放式厨房的设计拉近与顾客的距离。
The open kitchen is designed to get closer to the customers.
光带的设计放大了空间的纵深。
The design of the light ribbons has enlarged the depth of the space.
不同的包间力求统一格调，只在配饰中做些许改变。
The different VIP rooms strive for the unified style only with slight changes on the ornaments.

天街花园 Sky Street Garden

设计单位:杭州正午装饰设计有限公司 / 设计:陈石林、钟刚勇 / 面积:2200 m² / 主要材料:天堂鸟大理石、钛金不锈钢板、有影表格利夹板 / 坐落地点:杭州 / 工程造价:500万元 / 完工时间:2008年12月 / 摄影:陈杰

Design Unit:C+ZZ international design company limited / Design:Chen Shilin、Zhong Gangyong / Area:2200 m² / Main Materials:marble、stainless steel plate、plywood / Location:Hangzhou / Project Cost:RMB 5 000 000 / Finish Time:Dec,2008 / Photography:Chen jie

餐厅定位于商务、宴请、好友相聚等高端消费人群。设计摒弃一般人对高档餐厅“豪华”“繁琐”的感觉，改以简洁、大气的手法来营造空间，以大面积材质的搭配对比，突出材质本身的质感，并用精致的灯光来渲染，使整个餐厅透露出低调、奢华的感觉。为了获取最大的空间高度，设计师因地制宜地利用顶面隐蔽工程管道之间的空间，运用波浪形手法将整个高低不平的顶面贯穿起来，形成错落有致、优美而有趣的曲线，成为整个餐厅的亮点，让宾客留下深刻的印象。

This restaurant is orientated at the high-end consumers for business, feting and friend gathering etc. The design has thrown away the common feeling of ‘luxury” and “trivialness” on the superior restaurants and changed to create the space in a simple and grand way. The arrangement and contrast between the large areas of materials are applied to stress the quality of the materials. Furthermore, the delicate light is applied for decoration so that the whole restaurant is delivering a constrained and luxurious feeling. To achieve the maximum space height, the designer has properly made use of the space between the hidden pipelines on top and connected the whole rugged top surface by means of wave form to form a properly strewn, elegant and interesting curves, which has become a highlight of the whole restaurant and leaves deep impression on the guests.

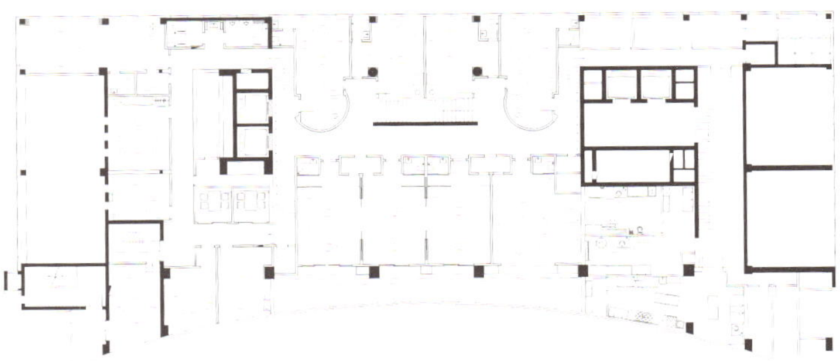

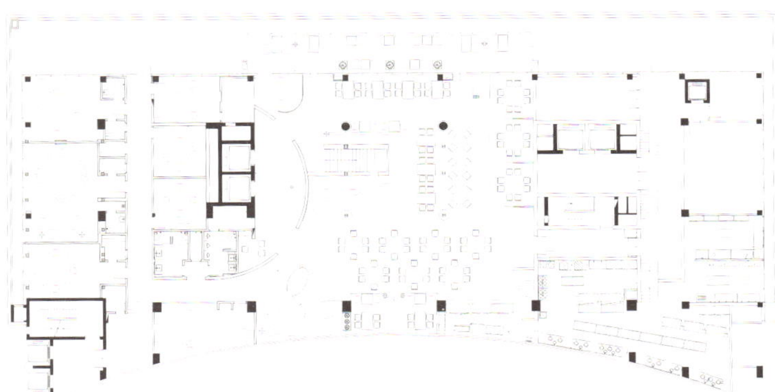

大面积材质的搭配对比，突出材质本身的质感。
The arrangement and contrast between the materials in large areas have stressed the qualities of the materials themselves.
运用波浪形手法将整个高低不平的顶面贯穿起来。
The technique in a wave form has connected the whole rugged ceiling together.

精致的灯光渲染，使包间透露出低调、奢华的感觉。
The exaggeration of the delicate light has endowed the VIP rooms with a low-profiled and luxurious atmosphere.
为了获取最大的空间高度，因地制宜地利用顶面隐蔽工程管道之间的空间。
To achieve the maximum space height, the space between the hidden pipe work in the ceiling are applied properly according to the actual conditions.
包间采光良好。
Good day-lighting in VIP room.
大堂餐厅全貌。
The full view of the lobby in the restaurant.

外婆家无锡店 The Grandma's Wuxi Branch

设计单位:杭州正午装饰设计咨询有限公司 / 设计:顾吉 / 面积:750 m² / 主要材料:木材、石材、红砖、涂料 / 坐落地点:无锡 / 工程造价:200万元 / 完工时间:2008年10月 / 摄影:贾方

Design Unit:C+ZZ international design company limited / Design:Gu ji / Area:750 m² / Main Materials: wood、stone、red brick、coating / Location:Wuxi / Project Cost:RMB 2 000 000 / Finish Time:Oct,2008 / Photography:Jia fang

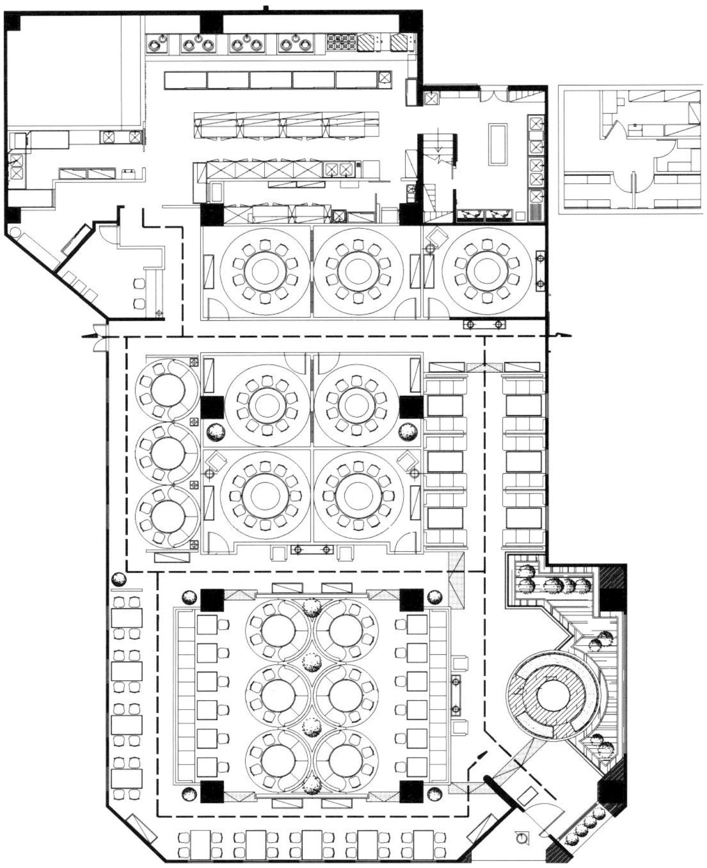

沉稳的色调既温馨又神秘。

The steady colors are both warm and mysterious.

餐厅的设计体现了浓郁的美式气息。

现代都市的人们渴望脱离钢筋丛林的压迫，更向往自然古朴。设计师为了让客人有一个放松愉快的用餐空间，采用木饰天花、铁艺吊灯、仿古石材，这些沉稳的色调贯穿整个餐厅的景色，当夜幕降临的时候，餐厅的灯光光束限定使这里的空间形成既温馨又神秘的气氛。

餐厅的每个立面，每张桌子，都制造出大气与和谐的气氛，红色砖墙与木饰反质家具的组合更显出整个气氛的品位与文化气质。独特的灯具，浓浓情郁的装饰画，讲究层次的布光，都是为了营造舒适的就餐环境。通透性的隔断，质地上乘的桌椅，更加显出用餐的你的精致品位。

The design of the restaurant has reflected a strong American style.

The modern metropolitans are desirous to break away from the oppression from the steel bar forest and more longing to the naturalness and simplicity. The designer has applied the wood decorated ceiling, iron artistic pendant lamp and the archaic stone materials to provide the guests with a relaxing and pleasant dining space. These steady colors have run through the scene of the whole restaurant. When the curtain of night falls, the light beams in the restaurant will confine this space into a warm and mysterious dining atmosphere.

Every facade and every table in the restaurant have created a grand and harmonious atmosphere. The combination of the red brick wall and wood veneered leather furniture has more set off the taste and cultural bearing of the whole atmosphere. The unique lighting fixtures, strong passionate decorative pictures as well as the daintily arranged light are all purposed to create a comfortable and secure dining environment. The transparent partition, high-quality tables and chairs have further shown your delicate dining taste.

餐厅的灯光光束限定使空间明暗相宜。

The light beams in the restaurant have defined the space in proper brightness and darkness.

通透性的隔断，质地上乘的桌椅，显出精致品位。

The transparent partitions and the superior tables and chairs have shown a delicate taste.

红色砖墙与木饰皮质家具的组合衬显整个气氛的品位与文化气质。

The combination between the red brick wall and the wood ornamented leather furniture has prominently set off the taste and cultural charm in the whole environment.

八府香鸭 Bafu Spiced Duck

设计单位:哈尔滨唯美源装饰设计有限公司 / 设计:韩冠恒 / 面积:8000 m^2 / 主要材料:瓷砖、岗岩、玻璃、油画大漆、锦缎 / 坐落地点:哈尔滨 / 工程造价:2400万元 / 完工时间:2008年8月 / 摄影:张奇永

Design Unit:WeiMeiYuan Design / Design:Han Guanheng / Area:8000 m^2 / Main Materials:ceramic tile、granite porphyry、glass、brocade / Location:Ha Erbin / Project Cost:RMB 24 000 000 / Finish Time: Aug,2008 / Photography:Zhang Qiyong

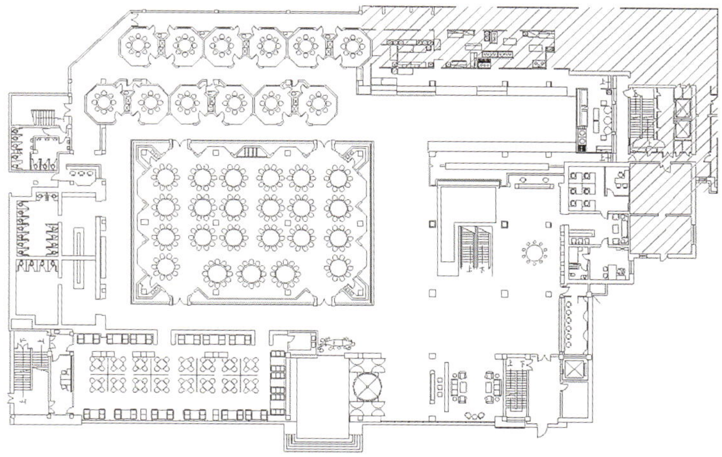

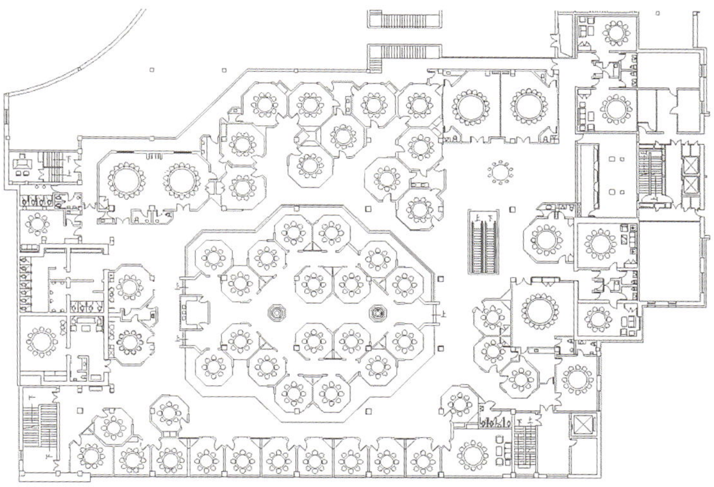

"八府香鸭"设计既由这家十几年历史的老店延展而来，形取之"八"，意取之"府"。形式上我们把房间都做成了八角形，这也是对餐饮行业的特殊性经过考查后得出的结论。即矩形的房间在餐饮包房中四个角都是实际使用功效不大的部分，而正八边形之间的平布组合可以产生出多种多样的变化。房间的面积被最大化利用后，余下的面积全部划给了公共空间。我们在充足的公共空间中作了三大水系，使一部分房间座落在水面之上，形成了室内的庭院建筑，又把现代化的交通工具滚梯也由水面而升，我们遵从主旨把清朝八大王府作简易冠名，也划分出较大的单体空间并作为装饰重点。家具及装置上又以传统手工艺"大漆"为主现场制作了超大尺度的家具，色彩上以红黑两中式传统颜色搭配，饰画上又以油画颜料来表现中国水墨写意荷花的内容。意在体现现代中式的主旨。外建上也以现代的材料与分割比例来演绎传统的符号与信息，来求得内外的统一，形与意的统一。

The design of "Bafu (Eight Mansions) Spiced Duck" is derived from this old restaurant with over ten years history, the structure from "Ba (Eight)" and the meaning from "Fu (Mansion)". Structurally octagonal rooms are designed to meet the particularity of catering industry. After our investigation, we reach the following conclusion that the four corners of rectangular rooms are not fully functioned in practical usage while plain combination of equilateral octagonal rooms can result in various changes. After the maximum usage of room area, the remaining area is all allocated to public space. Three water systems are designed in the public space. A section of rooms above the water appears to be interior courtyard construction while the rising of the escalator, a modern transportation means, from water surface, indicates that we follow our design idea of honoring the Eight Imperial Families of the Manchu. A large portion of single space is separated as the key place to decorate. Furniture and installation adopt traditional handicraft "lacquer" to produce super large scale furniture at site. Traditional color assortment of red and black are used in color designing while oil paint is used in decoration pictures to embody the lotus flowers in Chinese ink wash drawing, aiming to reflect the major idea of bearing modern Chinese style. In exterior design, modern materials and division proportion are applied to deduce the traditional signs and information in order to reach the unity between the interior and the exterior and the combination of the structure and the meaning.

一艘小船成就一处小景。
One small boat becomes a small scene.
不规则形状结构空间。
The space in irregular form and structure.

房间都做成了八角形。
The rooms are made octagonal.
红色隔栅分外抢眼。
The red grille is extremely eye-catching.
小包间静谧、优雅。
The small VIP rooms are quiet and elegant.
区隔而又通透的大空间。
The big divided but transparent space.

港式肥牛火锅 Hongkong Type Beef Hotpot

设计单位:飞形设计事业有限公司 / 设计:耿治国 / 参与设计:闫学江、马立荣 / 面积:290 m² / 主要材料:镜面不锈钢、绢丝壁纸、大理石、胡桃木、银箔漆 / 坐落地点:上海 / 完工时间:2008年4月 / 摄影:周宇贤

Design Unit:OFA Office for Flying Architecture / Design:Gustaf C.K.Kan / Participatory Design:Yan Xuejiang、Ma Lirong / Area:290 m² / Main Materials:mirror finished stainless steel、wallpaper、marble、walnut、paint / Location:Shanghai / Finish Time:Apr,2008 / Photography:Zhou Yuxian

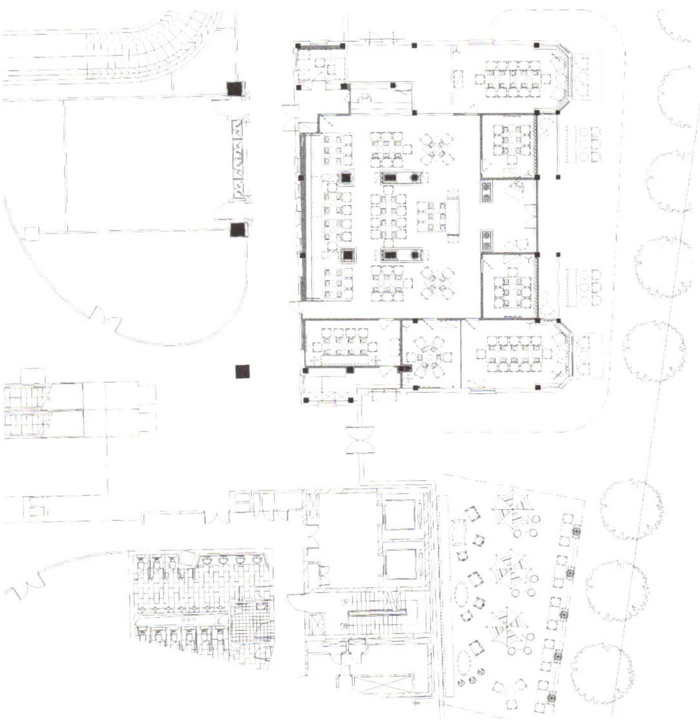

空间以金银色调统御，沿袭建筑外观的时代风格，在形成视觉冲击感的同时，将Art Deco引入室内，再现新艺术精神。水晶灯和柱列呼应，将人们带回20世纪二三十年代Art Deco正流行的时空中。

座椅和餐桌的选择在呈现精简线条的同时，又蕴含奢华感。墙面装饰复古奢华，还带着“书香门第”的高贵气，在自然与唯美路线中变幻出海派的高尚生活空间。饰有中国书法的墙面蕴藏着餐饮文化的内涵。

值得一提的是，设计师启用现代新式材质与工艺，对古典重新诠释。大厅柱列采用镜面玻璃制成，有一种炫丽的效果；正墙面包覆秋香绿绢丝手绘壁纸，展现香港维多利亚湾的繁华景象，呼应港式火锅餐厅主题；着意绘制的帆船与现代摩天楼群的对比，流露出历史时空的交错；人字纹胡桃木地板、灰色大理石唤起了旧日情怀。

The space is controlled by golden and silver colors and has continued the current style of the exterior building. It has introduced Art Deco into the indoor while forming a visual impact so that new artistic spirit is reproduced.

The selection of the chairs and dining tables has not only shown the concise and simple lines, but also contained the luxurious and magnificent affections. The wall decoration is antique and luxurious, but also has the nobleness of "a scholar's home". It shows a luxurious noble living space from the natural and aesthetic ways. The wall written with Chinese calligraphies has contained with rich meanings of catering culture.

The designer has applied the modern new materials and techniques to re-interpret the classism. The mirror glass is applied to the row of columns in the lobby, which has a flowery effect; the front wall surface is covered with the shadow green spun silk handwritten wall paper to show the busy scene of Victoria Harbor, Hongkong and echo with the subject of Hongkong Type Hotpot; the contrast between the deliberately drawn sailing boat and the modern skyscrapers has revealed an interlacement of historic time and space; the handmade lambdoid walnut wood floor boards and gray marble have evoked the reminiscent feelings.

火树银花的餐厅外观。

The glistering and shining appearance of the restaurant.

金银色调统御，延袭建筑外观的时代风格。

The space is controlled by golden and silver colors and has continued the current style of the exterior building.

墙面装饰复古奢华，带着“书香门第”的高贵气。

The wall decoration is antique and luxurious, but also has the nobleness of “a scholar’s home”.

水晶灯和柱列呼应，将人们带回20世纪二三十年代Art Deco正流行的时空中。

The crystal lamps echo with the column rows to bring the people back to the 20’s and 30’s of last century when Art Deco was popular.

现代品位 Modern Taste

设计单位:哈尔滨唯美源装饰设计有限公司 / 设计:张奇永 / 面积:2200 m^2 / 主要材料:石材、玻璃、饰面板、涂料 / 坐落地点:大连 / 工程造价:700万元 / 完工时间:2009年7月 / 摄影:张奇永

Design Unit:WeiMeiYuan Design / Design:Zhang Qiyong / Area:2200 m^2 / Main Materials:stone、glass、veneer 、coating / Location:Dalian / Project Cost:RMB 7 000 000 / Finish Time:Jul,2009 / Photography: Zhang Qiyong

在这个庞大的空间里，主要材质一目了然，大片的留白和建筑转角处的搭接与过渡又尽量突出对比感与整体感，无形中为空间注入一种恢宏却又细腻的气质。就好像为建筑内部穿上一件外衣：衣服的色彩内敛，款式简洁，却又通过细微的质地变化增强品位感。

In this huge space, the main materials are clear at a glance while the large area of blank wall as well as the joint and transition at the corners of the building have maximized the contrast and integrality, which has unconsciously infused the space with a magnificent but exquisite atmosphere. It is just like an overcoat for the interior building: its color is constrained and its style is simple while it has strengthened the taste through the tiny changes of the materials.

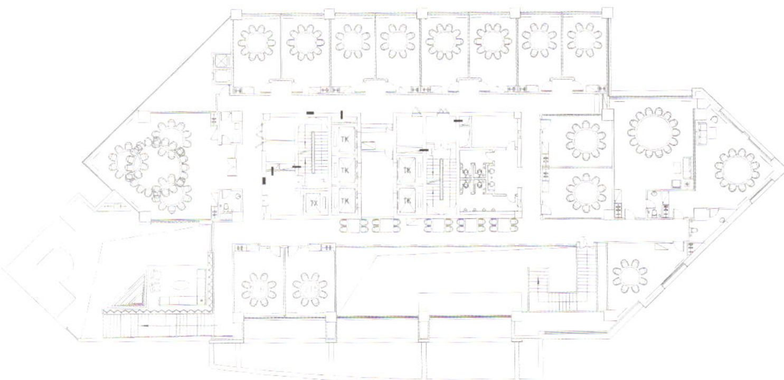

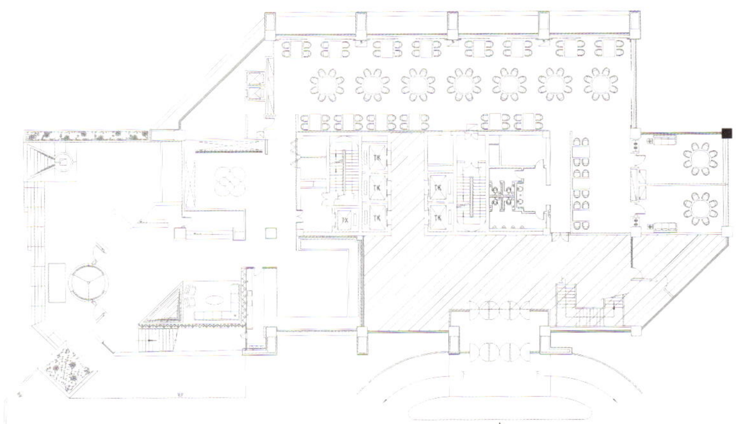

镜面的运用扩大了心理视觉感。
The application of the mirror has expanded the psychological vision.
交错排列的条状石材。
The strewn arranged striped stone materials.

黑白主打的空间秩序井然。
The space with black and white as main colors looks well-ordered.
端头的写真极具欧式古典韵味。
The portraitures at the end are of extremely continental-style classic charms.
餐厅局部。
Part of the dining room.

Mango Tree Mango Tree

设计单位:梁志天设计师有限公司 / 设计:梁志天、陈一峯、梁永贤 / 面积:508 m² / 主要材料:灰洞云石、云石、原木、皮革、柚木皮、古铜色镜子、墙纸 / 坐落地点:迪拜 / 完工时间:2008年6月 / 摄影:曾宪成

Design Unit:Steve Leung Designers Ltd. Architecture / Design:Steve Leung、Ashley Chan、Edmond Leung / Area:508 m² / Main Materials:marble、log、leather、teak、bronze mirror、wallpaper / Location: Dubai / Finish Time:Jun,2008 / Photography:Ulso Tsang

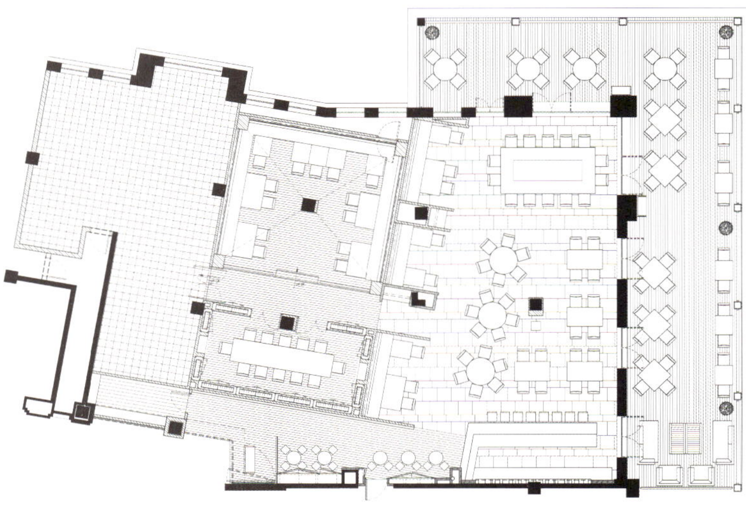

餐厅入口的柚木地板、铜镜饰墙及铁屏风，引领宾客来到酒吧区域。铁屏风以源自泰式建筑的长方形图案结合而成，由酒吧区域一直延伸到用餐区，远处是一盏耀眼的吊灯，把宾客的焦点带到用餐区域。

云石地台划分主用餐区，木珠组成的长形吊灯成为餐厅的另一焦点。吊灯下是一张由柚木及大理石制成的餐桌。餐厅的灯光由专人设计，既能于日间营造温馨的气氛，又可于晚间带来神秘的效果。餐厅饰以木及铁制的屏风，与入口的铁屏风呼应；啡色调子和由皮革及柚木制成的家具，加上东南亚风味的古董摆设，让每个角落洋溢着时尚又古雅的泰式风情。

藏酒库用上非洲崖豆木地板，凸显客人崇高的地位。大帐篷糅合现代室内建筑的神髓，特制灯笼映照金箔墙纸，极具大家气派；加上方形的座位设计，令这空间成为开小派对的闲适之地。

The teak floor boards, copper mirror decorated wall as well as iron screen at the entrance of the restaurant have led the customers to the bar area. The iron screen is assembled by the square patterns from Thai-style buildings and extends from the bar area to the dining area. The distant glistening pendant lamp has brought the focuses of the customers to the dining area.

The marble platform divides the main dining area while the long pendant lamp assembled by the wood beads becomes another focus of the restaurant. Under the pendant lamp is a dining table made of teak and marble. The light in the restaurant is specially designed, both for the creation of a warm atmosphere at the daytime and for the introduction of a mysterious effect at nighttime. The restaurant has been decorated with the wood and iron screens, echoing with the iron screen at the entrance. Coffee tones as well as the leather and teak furniture together with the display of the antiques in Southeast Asian style has made every corner overflow with a fashionable, classic and elegant Thai-style atmosphere.

African Wenge floor boards are used in the cellar to prominently show the superior position of the customers. The marquees have mixed with the essences of the modern interior architecture. The specially made lanterns shine upon the golden foil wall paper, which extremely has a master's manner; in addition to the design of the square seats, this space becomes a leisurely and comfortable place for small parties.

餐厅的灯光由专人设计，既能于日间营造温馨的气氛，又可于晚间带来神秘的效果。

The light in the restaurant is specially designed, which not only can be used to create a warm atmosphere at daytime, but also bring about the mysterious effect at nighttime.

木珠组成的长形吊灯成为餐厅的另一焦点。

The long pendant lamp constituted by wood beads becomes another focus in the restaurant.

藏酒库用上非洲崖豆木地板，凸显客人崇高的地位。

African Wenge floor boards are used in the cellar to prominently show the superior position of the customers.

每个角落洋溢着时尚又古雅的泰式风情。
Every corner is overflowed with a fashionable, classic and elegant Thai-style atmosphere.
啡色调子和由皮革及柚木制成的家具。
Coffee furniture as well as the leather and teak made furniture.
东南亚风味的古董摆设。
The antiques and displays in Southeast Asian style.

唐会·悦雍轩 Tang·Yue Yong Xuan

设计单位:江苏省海岳酒店设计顾问有限公司 / 设计:姜湘岳、吴海燕 / 参与设计:徐云春、卞玲、张逸轩 / 面积:2000 m^2 / 主要材料:日本墙纸、石材、布艺 / 坐落地点:南京 / 工程造价:1200万元 / 完工时间:2009年1月 / 摄影:莫尚勤

Design Unit:Jiangsu Haiyue Hotel Design Consulting Co., Ltd. / Design:Jiang Xiangyue、Wu Haiyan / Participatory Design:Xu Yunchun、Bian ling、Zhang Yixuan / Area:2000 m^2 / Main Materials:wallpaper、stone、cloth / Location:NanJing / Project Cost:RMB 12 000 000 / Finish Time:Jan,2009 / Photography: Steve Mok

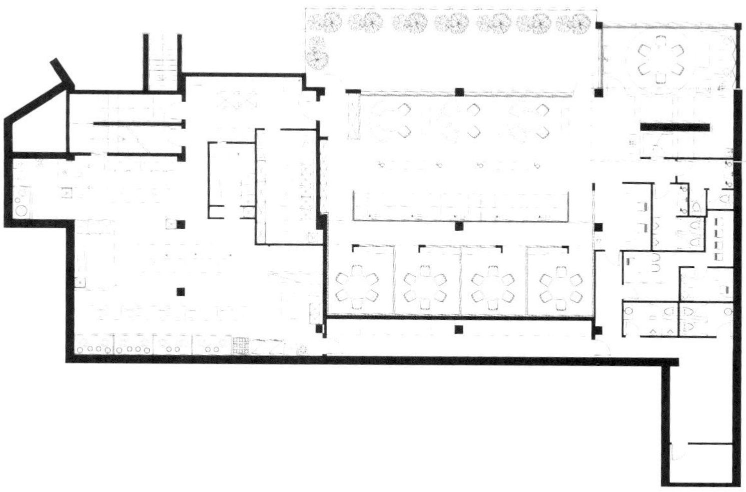

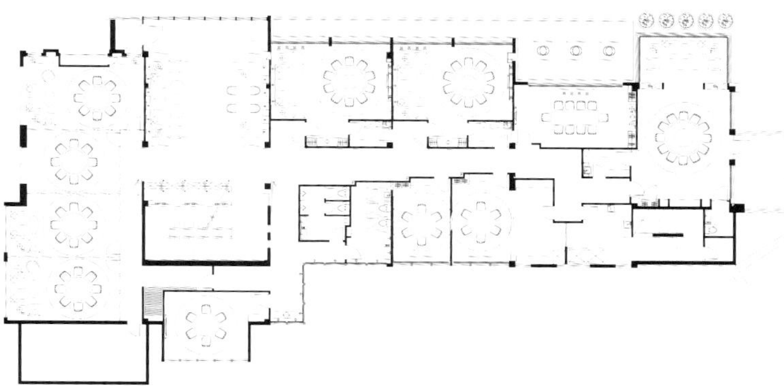

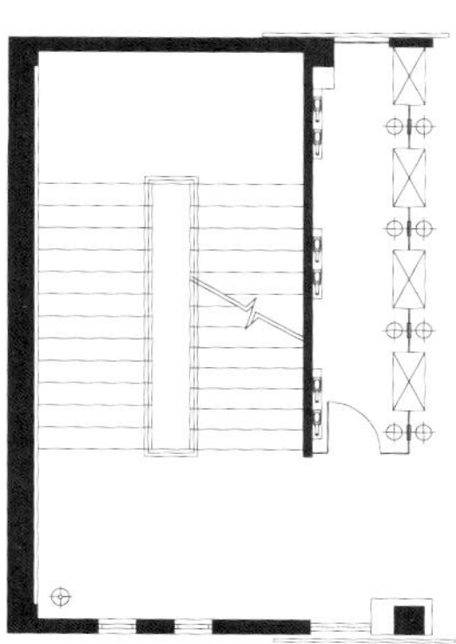

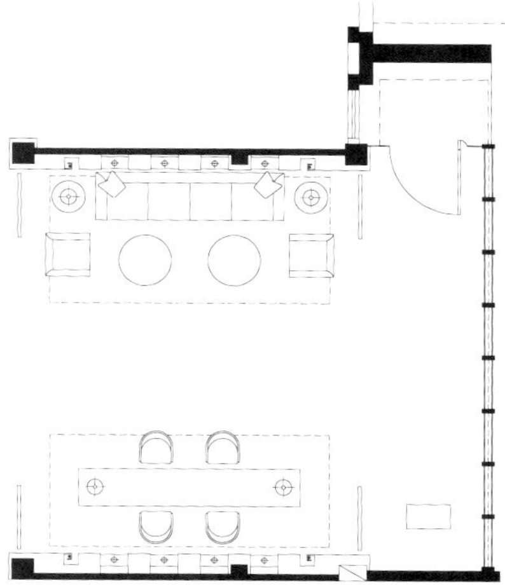

设计中，将中国古典主义建筑和现代主义相结合。大堂迎宾处，精心挑选的迎宾花与背景形成良好的对比状态。顶和地之间大体块的结构感、接待台背后的古书对空间进行了柔化。走道采用古典主义矮墙窗棂的方式，用镜面加木格的做法，使客人能够在木格的序列中看到周围的影像。

中餐包间的顶部、墙面的深色，把属于中式古典主义的情怀用现代的方式表现出来，体现出高贵感。特殊的超长无缝玻璃窗让美景尽收眼底。

豪华包间极具书香门第的气氛。在餐厅里面放置了大量的古典主义书籍，使得空间充满了浓浓的书香气味。古典主义的建筑采用现代主义的表现方式，并且采用深色的木饰面，使得空间非常具有高贵感。

中餐零点空间的设计采用通灵的方式，使顶面的吊灯产生一个良好的倒影关系，减弱了空间层高较矮的压抑感。中间的案台更好地与功能结合，既能做案台又能做备餐台，解决了零点过程中服务员的实际问题。由于半封闭的零点包厢在黑区，故采用发光玻璃的设计方式，使得空间很透明，减少客人的压抑感。

楼梯的处理使得空间更具私密感，设计时不用通透的玻璃，以免上下的客人之间由于对视而产生尴尬的感觉。深色的走廊里面摆放了很原始的艺术品，是由专门的艺术家进行创作的，艺术品是整个空间的灵魂。

原本被业主计划用作仓库的房间是一个建筑条件十分差的空间，然而设计却让这个空间最终成为了一个亮点。在顶部开启了天窗，木饰面选用了浅色，整个空间显得很通透，视线往前往上都能看到美景，把整个建筑的劣势转为优势，使包间非常具有吸引力。包间外坡屋顶较矮处的地面，采用地面灯的方式，提醒客人层高较低，当心碰到，同时表达出低矮处回转的一种情感。

his building was formerly used for block management. The owner rented it due to its geographic location. During the design, the building has been reshaped to finally realize the proper combination between the interior and exterior.

The design has combined the Chinese classic architecture with the modernism. At the reception hall in the lobby, the usher flowers have formed a good contrast with the background while the ancient books behind the reception desk have softened the space. The classic low-wall muntin is applied to the corridor and the guests can see the ambient images through the serial wood grids by way of the mirror in combination with the wood grid.

The ceiling in the VIP rooms for Chinese cuisines and the deep color of the wall surface has demonstrated the Chinese classic moods through the modern methods. The luxurious VIP rooms have the extremely literary atmosphere. The classic architecture has applied the modernism ways to make the space very noble and dignified. The zero space for the Chinese cuisines has been designed in a supernatural way so that the pendant lamps in the ceiling has created an appropriate inverted reflecting relation and weakened the oppression of the quite low floor height. The room that was used as a warehouse by the owner is a space of very bad architectural conditions, but the design has made it a shining point finally. The ceiling is opened with a skylight and light wood veneers are chosen so that the whole space looks transparent and you can see a beautiful scenery when you raise you head up. The disadvantages are changed to the advantages so that the VIP rooms become very attractive. The outdoor sloping roof of the VIP rooms has been applied with the floor lamps in its comparatively lower floor to remind the guests of the low floor height in case of bumping their heads, which also have reflected a changed feeling at the low position.

中式古典主义的情怀用现代的方式表现。

The Chinese classic moods are demonstrated through the modern methods.

中餐零点空间的设计采用通灵的方式，使顶面的吊灯产生一个良好的倒影关系。

The individual ordering space for the Chinese cuisines has been designed in a supernatural way so that the pendant lamps in the ceiling have created an appropriate inverted reflecting relationship.

顶部开启了天窗，木饰面选用了浅色，整个空间显得很通透。

The ceiling is opened with a skylight and the wood veneers are selected with light color instead of the dark in the other spaces so that the whole space looks transparent.

深色的木饰面以及墙面上的仕女图形成一种古典主义高贵的感觉。

The dark wood veneers and the beauty pictures on the wall deliver a classic noble atmosphere.

大堂迎宾处精心挑选了迎宾花与背后的背景形成非常良好的对比状态。

At the reception hall in the lobby, the elaborately selected usher flowers have formed a good contrast with the background in the back.

一茶一坐 Chamate

设计单位:无锡市上瑞元筑设计制作有限公司 / 设计:曹小鸥、冯嘉云 / 面积:650 m² / 主要材料:仿古砖、饰面板、原木、玻璃 / 坐落地点:无锡 / 工程造价:90万元 / 完工时间:2008年12月 / 摄影:文宗博

Design Unit:Shang Rui Yuan Zhu Design Co., Ltd. / Design:Cao Xiaoou、Feng Jiayun / Area:650 m² / Main Materials:rustic tile、veneer、log、glass / Location:Wuxi / Project Cost:RMB 900 000 / Finish Time: Dec,2008 / Photography:Wen Zongbo

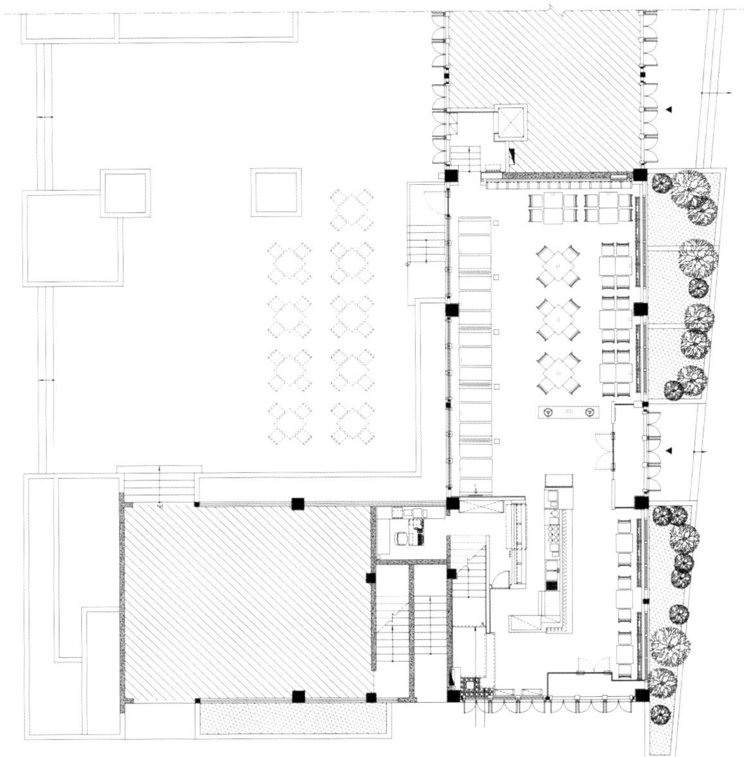

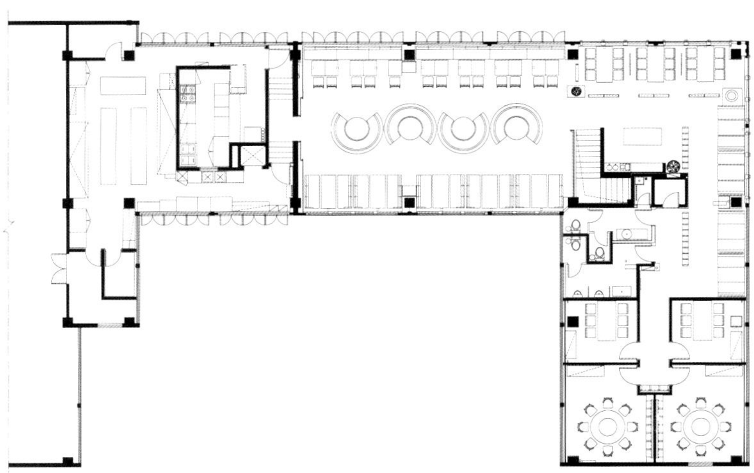

古朴的质感，沉着温和的色调，让人心境顿然轻松安逸。
The simple and unsophisticated quality as well as the steady and warm colors will relax and comfort you immediately.

一茶一坐，将中国传统茶文化和中国特色美食融合一体的全新时尚餐饮连锁店，将传统文化成功植入现代人士的时尚生活中，不仅提供传统意义上的味觉享受，更体现一种对精致生活的质量追求。

很巧合，这一次的店面选址刚好位于南长古街的入口处，建筑是一幢古旧的中式楼房，有一块与老街相通的后庭院，很有江南别院的风情，跟饮茶的意境相当契合；大家都有意于借这一份情景，因此就有了这样一家一茶一坐的私家别院。

外观保留并延续了房屋的大部分构造和语言，只将与后庭院相隔的立面整体改造，结合老建筑的风格，形成敞开的玻璃长窗，把室外的街景、阳光引入房间，再加一道帘，从内往外看，尽收眼底，闹中取静；从外往内望，隐约含蓄，让人期待。转到室内，古朴的质感，沉着温和的色调，让人心境顿然轻松安逸。坐下，点一壶茶，起身，再细细玩味精心收藏的老家具、看似不经意的玩件摆设，一灯、一座恰如其分。踱到二楼，此处的风景引用了Vergangene Weiten的版画作品，私自将画中的风景借过来，用在这私家别院里，虚虚实实，西法中用，相得益彰，幽默，却值得细细赏玩。约三五好友饮茶聚餐海阔天空，似是自家宅院，怡然自得。

The ship is located at the entrance of Old Nanchang Street. The building is an antique old Chinese-style building with a backyard connected with the old street. It has quite an atmosphere of a courtyard in the south of Yangtze River and quite matches the artistic conception of drinking teas; everyone intentionally borrows such scenery and thus come to this private courtyard.

Its appearance has been reserved and most of the structure and language have continued. Only the facade separated from the backyard is entirely modified. In combination with the style of the old building, it has formed into an opening long glass window. With one curtain, it looks concealed and connotative in a view from the inside to the outside. Turning to the interior, its simple and antique quality sense and warm colors suddenly relax you while the delicately collected old furniture and the seemingly casual antiques and furnishings are appropriate in their way: one light, one seat; further to the second floor, the woodcuts by Vergangene Weiten are introduced to this private courtyard, which look appropriate to each other by applying the western way to the Chinese use. You can gather with several friends to drink tea, have dinner and chat freely, just like in your own courtyard, joyful and contented.

一灯、一座恰如其分。
One lamp, one seat, just appropriate.
精心收藏的老家具成为空间的主角。
The elaborately collected old furniture has become the key role in the space.
保留并延续了房屋的大部分构造和语言。
Keep and continue most of the building structure and language.
看似不经意的玩件摆设值得细细赏玩。
The seemingly casual displays are worth carful enjoyment.

尝香汇餐厅 Changxianghui Restaurant

设计:王善祥 / 参与设计:王善辉 / 面积:1218 m² / 主要材料:瓷砖、钢管、钢板、竹面板、玻璃 / 坐落地点:上海 / 工程造价:200万元 / 完工时间:2008年10月 / 摄影:老猫

Design:Wang Shanxiang / Participatory Design:Wang Shanghui / Area:1218 m² / Main Materials:ceramic tile、steel tube、steel plate、bamboo face plate、glass / Location:Shanghai / Project Cost:RMB 2 000 000 / Finish Time:Oct,2008 / Photography:Lao mao

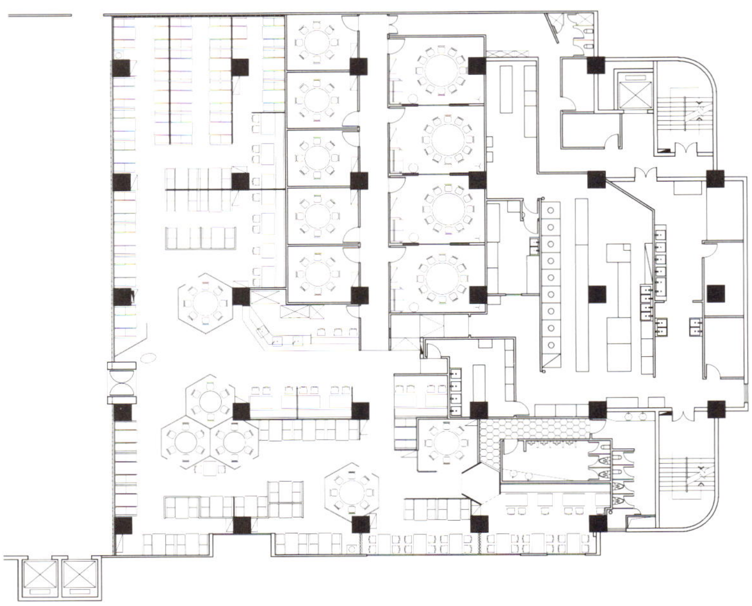

对着大堂的一面玻璃墙全部用钢管竖向排列成格栅。

The glass wall opposite to the lobby is lined into the grids by the steel pipes.

六边形小格子让小包间与大空间欲隔实合。

The small hexagonal grids make the small VIP rooms and big space seemingly separated, but actually connected.

如何把餐厅包装成一家独具个性的场所是设计师首要考虑的问题，吸引以年轻人为主的客人。设计师把餐厅想象为一个充满六角形的蜂巢，六角的蜂巢形既是自然的造化，又是个典型的充满了人造痕迹的工业造型，蜂巢数量的无限也象征了业主的餐厅业务扩展的无限前景。

店址选在一栋商务楼的4层。由观光电梯上楼后，首先进入一个150m²的大堂，司一家咖啡馆和游戏机房共用。由于游戏机房进出的多是些年轻人，难免吵闹，因而对着大堂的一面玻璃墙全部用钢管竖向排列成格栅，上面高低错落地开一些六角形窗口，构成看与被看的方式。进门几个六角形平面的隔间与几面隔墙分割、围合出“庭院深深深几许”的趣味空间。隔间和隔墙立面上大小不同的六角形开口营造视觉上透而不透，空间上隔而不隔的层次，墙上、地面及天花都有不同尺寸与材质的蜂巢形出现，体现浓郁的现代氛围。

整体色彩以白色为主，配以苹果绿，搭配米色竹板饰面。黑色餐椅椅背侧边分别使用红、黄、绿等小条鲜艳色彩，使整个空间的色彩活跃起来，渲染亮丽的青春气息。

What a designer shall take first consideration into is how to decorate a restaurant into a unique individualized place to mainly attract the young customers. First the designer has imaged the restaurant as a honeycomb full of hexagons. The infinite quantity of the honeycombs also symbolizes the infinite prospect of the business development of restaurant.

The restaurant is located at the fourth floor of a business building. To avoid the noises, the glass wall opposite to the lobby is lined up by the steel pipes with some hexagonal windows on them, which forms a way of watching and being watched. The separations at the entrance and the partition walls have be divided into and enclosed an interesting space of “profound courtyard”. The hexagonal openings in different sizes on the wall have created a transparent but not penetrative level sense. There are also honeycomb patterns on the wall, ground and ceiling to reflect a strong modern atmosphere.

The entire color is mainly white, arranged with apple green and the beige bamboo veneer. The stripped bright colors such as red, yellow and green are separately used on the side of the black dining chair back so that the colors in the whole space get activated to render a bright adolescent atmosphere.

几面隔墙分割，围合出“庭院深深深几许”的趣味空间。

Seven partition walls have divided into and enclosed an interesting space of “profound courtyard”.

隔墙立面上大小不同的六角形开口营造视觉上透而不透，空间上隔而不隔的层次。

The hexagonal openings in different sizes on the wall have created a visually transparent but not penetrative level and look like divisions in space.

整体色彩以白色为主，配以苹果绿，搭配米色竹板饰面。

The entire color is mainly white, arranged with apple green and the beige bamboo veneer.

超级牛扒 SUPER STEAK

设计:吴纯 / 面积:1500 m² / 主要材料:镀钛不锈钢、金镜、茶镜、皮革、山西黑花岗岩、花梨木皮、青砖 / 坐落地点:苏州 / 工程造价:500万元 / 完工时间:2009年3月/摄影:文宗博、潘宇峰

Design:Wu chun / Area:1500 m² / Main Materials:stainless steel、gold mirror、solar mirror、leather、black granite、rosewood、blue brick / Location:Suzhou / Project Cost:RMB 5 000 000 / Finish Time: Mar,2009 / Photography:Wen Zongbo、Pan Yufeng

设计师以前卫、时尚的风格与本土元素混搭，赋予餐厅独特的魅力。

整个餐厅区域划分明朗，主要由开放式用餐区和包厢区两部分组成。开放式用餐区面积较大，布局上采用敞开式厨房和自助式餐吧的形式，餐位布置紧凑却不显局促。以深色为基调，橙红色为点缀，配合独特的灯光设计营造浪漫的氛围。灯光在夜晚被调暗，重点照明只汇聚于餐桌上。顶部矩阵排列的方块茶色镜面，令开放式的大空间在感观上又非常有序，并通过对下部的反射丰富空间层次。开放式用餐区和包厢区通过半透的陈酒柜区隔，花梨木饰面的陈酒柜设计灵感来自园林的花窗和传统博古架，以现代构成的手法组合；大厅的立柱贴上青砖及砖雕，在外面罩上有色的透明夹胶玻璃，展现时尚气息；包厢区设计风格贵气、儒雅。爵士白大理石的优美纹理，深棕色皮革的高贵质感，古铜色镀钛镜面不锈钢的精致华美，酸洗古铜镜雕刻有中国传统祥云流水纹样透出的几许尊贵……诸如此类，无不诠释出传统与现代、历史与时尚的交融。

The designer has mixed the pioneering and fashionable style with the local elements to endow the restaurant with the unique charms.

The whole restaurant has a clear division among the areas, mainly consisting of the opening dining area and VIP room area. The opening dining area has quite a large area with arrangement of the opening kitchen and self-served food bar. It mainly applies dark colors dotted with salmon pink and creates a romantic atmosphere together with the unique light design. The top square brown mirror in a matrix has made the big opening space visually in a good order and also enriches the space levels through the reflection of the lower parts. The opening dining area and VIP room area are divided by the translucent wine cabinet. The design of the rosewood-veneered wine cabinet is inspired by the lattice windows in the gardens and traditional antique shelves, which are combined together in the modern constitutive technique; the vertical columns in the lobby are pasted with grey and carved bricks and covered with the colored transparent laminated glass to deliver a fashionable atmosphere. The VIP room area is designed noble and cultured. The jazzy white marble, nut-brown leather, bronze-colored titanium coated stainless steel as wel as pickled antique bronze mirror carvings all have demonstrated an integration between the tradition and modernity as well as history and fashion.

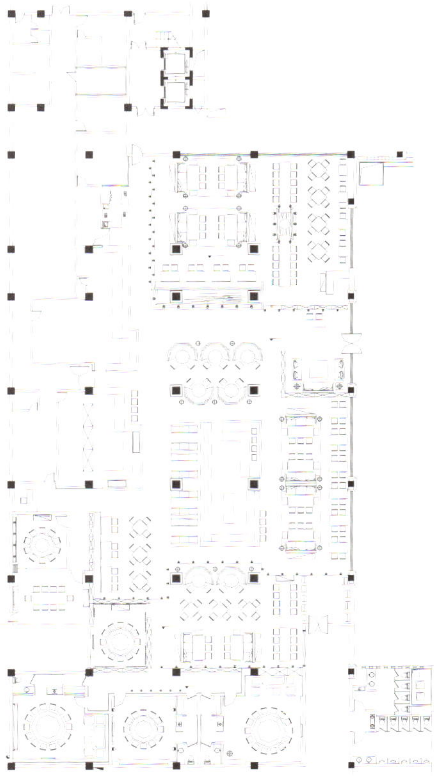

餐位布置紧凑却不显局促。
The dining seats are arranged compact without constraints.
镜面通过对下部的反射丰富空间层次。
The mirror has enriched the space levels through the reflection of the lower parts.

A座

V07

顶部矩阵排列的方块茶色镜面，令开放式的大空间在感观上又非常有序。

The top square brown mirror in a matrix has made the big opening space visually well-ordered.

以深色为基调，橙红色为点缀，配合独特的灯光设计营造浪漫的氛围。

It mainly applies dark colors dotted with salmon pink and creates a romantic atmosphere together with the unique light design.

灯光在夜晚被调暗，重点照明只汇聚于餐桌上。

The light is adjusted dim at night and the key lighting is focused on the dining tables.

包厢区设计风格贵气、儒雅。
The VIP room area is designed noble and cultured.
深棕色皮革尽显高贵质感。
The dark brown leather has fully revealed a noble charm.

德福时尚火锅 Defu Fashion Hot-Pot Restaurant

设计单位:佳木斯市豪思环境艺术顾问设计公司 / 设计:王严民 / 面积:900 m^2 / 主要材料:玻化砖、曲柳擦色贴面板、木格、油漆、壁纸、布艺、工艺玻璃、涂料 / 坐落地点:鹤岗 / 工程造价:120万元 / 完工时间:2008年9月 / 摄影:王严民

Design Unit:HaoSi Circumstance Artistic Consultant Design Corporation / Design:Wang Yanmin / Area:900 m^2 / Main Materials:glazed tiles、face plate、paint、wall paper、technology glass、coating / Location: Hegang / Project Cost:RMB 1 200 000 / Finish Time:Sep,2008 / Photography:Wang Yanmin

德福时尚火锅以宴请的消费方式和包房的格局进行设计，空间调性主要运用稳重的黑色，借此营造用餐氛围的低调与华丽，适度融合激情的红色，点睛之处寓意着现代中式的美感写意，同时将一些中式元素符号以简约的手法表现出来，洗练、利落中备显空间震撼、空灵，让客人在消费时内心感受着一份平和与舒畅。

The restaurant is designed for banquets in the form of compartments. Sober black is applied to display simplicity and elegance of the dinner, with passionate red colors combined moderately, indicating the modern Chinese-style beauty. Some Chinese elements are also displayed in a simple style, fully displaying the imp-essive and intangible space in a refined way, to bring peaceful mind to customers.

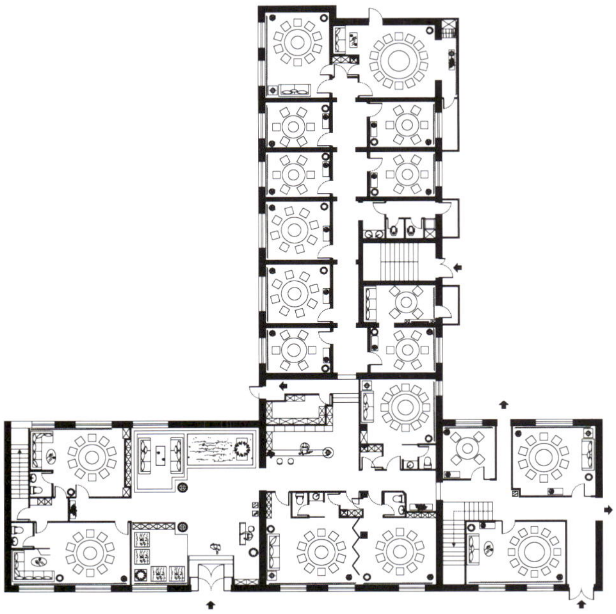

店名被意象为印章，饰以激情的红色。
The shop name is a symbol of stamp decorated in fire red.
空间调性主要运用稳重的黑色。
The space theme is the steady black.
明暗有致的光影把控。
Well arranged light and shade.

中式元素符号以简约的手法表现出来。

Chinese elements presented by simple means.

现代中式的美感写意。

Modern and Chinese style beautiful expression.

包间一角。

View of the compartment.

洗练、利落中备显空间韵味。

Space charm enriched by simple and spiffy designs.

蜂蝶来西洋会馆 Fengdielai Western-Style Exclusive Restaurant

设计单位:佳木斯市豪思环境艺术顾问设计公司 / 设计:王严钧 / 面积:1200 m² / 主要材料:水磨石、马赛克、黑檀板、黑钢板、玻璃、布艺、壁纸、乳胶漆 / 坐落地点:鹤岗 / 工程造价:200万元 / 完工时间:2008年11月 / 摄影:王严钧

Design Unit:HaoSi Circumstance Artistic Consultant Design Corporation / Design:Wang Yanjun / Area:1200 m² / Main Materials:terrazzo、mosaic、black steel plate、glass、cloth、wallpaper、emulsion paint / Location:Hegang / Project Cost:RMB 2 000 000 / Finish Time:Nov,2008 / Photography:Wang Yanjun

作为新式时尚用餐场所，古典、现代、时尚、华贵是本案的中心构想，设计手段克服内建筑物的低矮空间，利用镜面反射与水晶玻璃、LED光影荡漾水晶吊灯，凸显晶莹剔透的空间氛围，软包、布艺、皮革、马赛克营造了雅致低调的华贵。

Designed for a new fashion restaurant, the scheme focuses on classics, modernity, fashion and luxury. The design is not limited to the low space of the inner building, but reflects the clear and crystal space with mirror reflection, crystal glass, and LED crystal pendant lamps, while creates elegant luxurious atmosphere by upholstered furniture, fabric decoration, leather, and mosaic.

会馆前台。
Reception desk of the club.
蜂巢型元素贯穿了顶面和服务台。
Honeycomb-like elements distributed on both the top and the service desk.
LED光影荡漾水晶吊灯。
LED shadow under the crystal pendant.

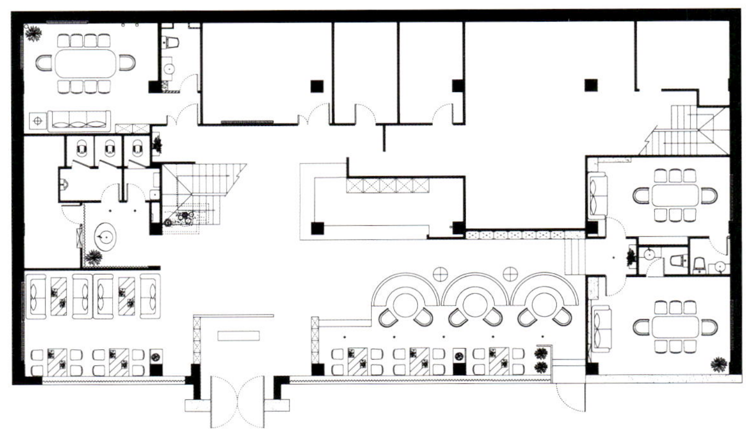

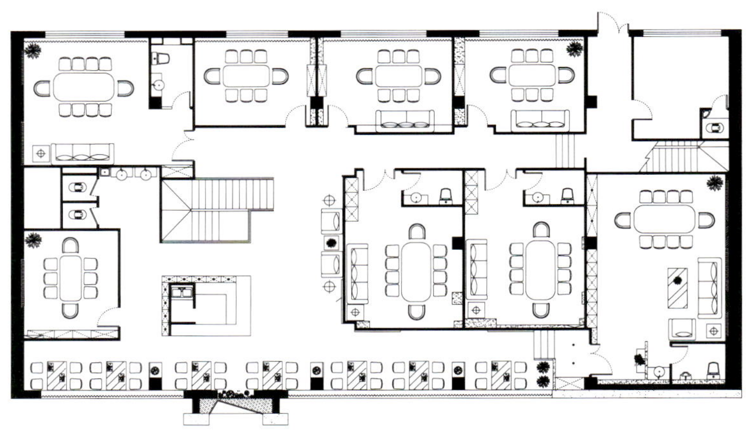

利用镜面反射丰富空间层次。
Space gradation enriched by the mirror reflection.
设计手段克服内建筑物的低矮空间。
Low space of the internal units overcome by special designs.
软包、布艺、皮革、马赛克营造了雅致低调的华贵。
The elegant and low-key luxury is created by soft package, cloth art, leather and mosaic.

港丽餐厅（龙之梦店） Charme Restaurant

设计单位:杨旭建筑工作室 / 设计:杨旭、白乐 / 面积: 960 m² / 主要材料:木、烤漆玻璃、铝板 / 坐落地点:上海 / 完工时间:2009年1月 / 摄影:刘圣辉

Design Unit:YangXu Architecture Studio / Design:Yang xu、Bai le / Area:960 m² / Main Materials:wood、paint coated glass、aluminium panel / Location:Shanghai / Finish Time:Jan,2009 / Photography:Liu Shenghui

港丽餐厅在设计时希望突破一般餐厅那种敞开式的大堂布局，敞开就意味着会有很多人拥挤在同一个空间中，声音不免嘈杂。镂空木隔断制成的飘带游刃于偌大的空间中，既具有形式美感，又丰富了空间层次，打破一览无余的开敞。为了让来此就餐的人们有不同的体验，餐厅的座位形式多变，卡座、常规座、圈座交替穿插，赋予餐厅更多生趣。包间的墙面不仅仅作为隔断，更是一种空间主体元素的表达。

Hoping overcome the common lobby layout that is open-style, Charme Restaurant is designed to many independent spaces and a quiet environment. The hollowed-out wooden partition in the huge spaces is full of beauty and rich of spatial levels which break the glance of the wide space. In order to fulfill people' s different taste, the restaurant' s form varied with square seats, conventional seats and ring seats which bring joyfulness to the restaurant. The wall of the private room become a partition part and also an element of the space theme.

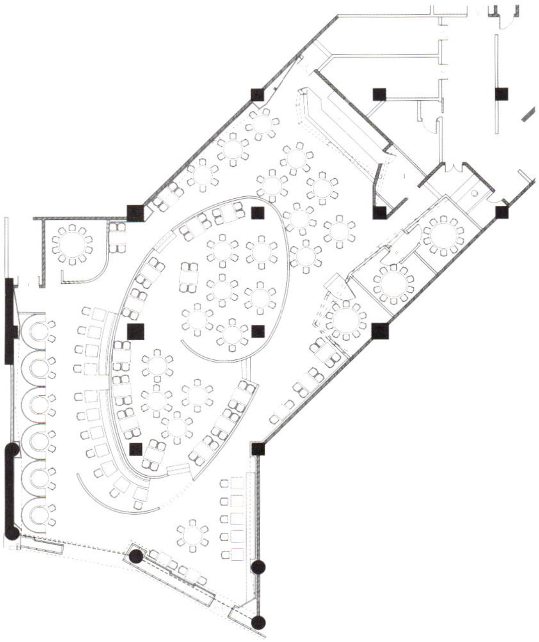

镂空木隔断丰富空间层次。
Hollowed-out wooden partition brings rich spatial level.
敞开和半封闭只在一念之间。
Open and semi-closed just between a moment in time.
特写镂空木隔断。
Close up to the hollowed-out wooden partition.

木隔断即是亮眼的装饰元素，也是包间墙面。

The hollowed-out wooden partition is a bright decorating element, and a private room wall as well.

光面和糙面的结合。

The combination of smooth and rough surfaces.

不同的座位形式齐聚一堂。

Different forms of seats gathered together.

倚墙而建的弧形餐厅。

The arc restaurant against the wall.

名湖美景大酒店 Minghu Meijing Grand Hotel

设计单位:南京广暄工程设计顾问有限公司 / 设计:陈广暄、马珺 / 面积:1500 m² / 主要材料:木纹复合板、仿皮纹壁布、古铜镜钢、双色拼花饰面板、黑色工艺镜 / 坐落地点:南京 / 工程造价:450万元 / 完工时间:2009年3月 / 摄影:南京远景摄影设计有限公司

Design Unit:Nanjing Guangxuan Design / Design:Chen Guangxuan、Ma Jun / Area:1500 m² / Main Materials:composite board、wall hanging、bronze mirror、veneer、technics mirror / Location:Nanjing / Project Cost:RMB 4 500 000 / Finish Time:Mar,2009 / Photography:Nanjing YuanJing photograph Design Co., Ltd.

本案是对原酒店的一次扩建，业主旨以全新的风格创造一个更适合高端客群的消费场所。为了让新老项目能够形成风格的过渡，保留了老项目中的部分图形元素，放弃了原有显性的色彩搭配，而以更多隐性的细节设计来提升空间品质。暖灰色的皮纹壁布配合大块面的黑色工艺镜加之蜡烛吊灯营造出一种雅致格调。“鱼”于“水”的形象渗透在整个装饰设计中。包间入口有二维图形设计的水波纹门扇和三维构成设计的水波纹壁饰。走廊橱窗里有全玻璃吹制的水草和鱼群。这些细节虽没有对风月文化的追逐与膜拜，却能带给人一份对自然情趣的隐约联想与淡淡感动。

It is an expansion to the existing hotel. The owner is purposed to create a consumption place more suitable for the high-end customer groups in a brand-new style. To make the old and new project form a transition of style, the designer has reserved the partial figural elements in the old projects. He has given up all the existing visual color arrangements and applied more indistinct design on the details to upgrade the space quality. The warm gray leather texture wall cloth in combination with the large area of black art mirror together with the candle pendant lamp has created an elegant atmosphere. The image of "fish" in "water" has run through all the decorative design. There are the water wave door leaves designed by the two-dimensional figure and the water wave wall ornaments designed by three-dimensional figures at the entrance to the VIP rooms. There are water grasses and fish clusters completely blown by glass in the shop windows in the corridor. Although these details have no pursuit and worship to the romantic culture, they can bring a faint association to the natural affections and a faint emotional touch to people.

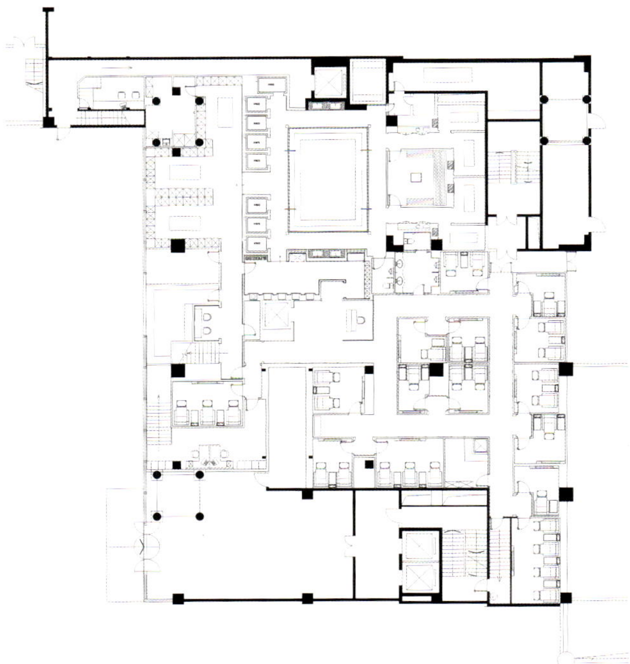

生动有趣的细部。
The vivid and interesting details.
造型别致的吊灯。
The pendant lamp in unique model.

Arletta

黑色茶几搭配白色沙发。

The black tea table is arranged with the white sofa.

玻璃与镜面的使用，也围绕着扩大空间与让空间之间有更多交流而进行。

The application of glass and mirror is also centered on the expansion of space so that there can be more communications among the spaces.

糙面的背景墙在灯光的辉映下极具肌理感。

The roughened background wall looks extremely textured under the reflection of the light.

卧室一角。

One corner of the bedroom.

通透灵动的卫生间。

The transparent and bright washroom.

荷花的浪漫 Romance of Lotus

设计单位:汕头市蓝鲸室内设计有限公司 / 设计:陈骏 / 面积:144m² / 主要材料:半抛光砖、茶镜、微晶白人造石、木饰通花、复合木地板、墙纸、不锈钢、油画 / 坐落地点:汕头 / 工程造价:30万元 / 完工时间:2008年11月 / 摄影:邱小雄

Design Unit:Shantou LanJing interior design corporation, Ltd. / Design:Chen jun / Area:144m² / Main Materials:polished tiles、tawny glasses、artificial stone、wood veneer、compound wood floor、wallpaper、stainless steel、 painting / Location:Shantou / Project Cost:RMB 300 000 / Finish Time: Nov,2008 / Photography:Qiu Xiaoxiong

荷，大约是中国人一种人文精神寄托；浪漫，大约是年轻一代的生活方式。

设计师在本案中凭借对生活、人文的感悟，于空间拓展上着墨，以“步移景动”的设计手法，将餐厅作为空间重点及人流动线的交汇处，在空间布局上对餐厅进行视觉拓展，运用玻璃、茶镜等材质去做空间分隔，同时考虑了餐厅的采光需求。

浪漫是需要色彩与时尚氛围的，所以有如水晶般透明的玻璃卫浴间；所以有温馨的色调，时尚的摆件和家具。荷花这一人文主题贯穿整个餐厅设计，并将其安排在餐厅背景墙中央，成为家中最为亮丽的色彩及视觉焦点，演绎着“很中国的”荷花浪漫。

Lotus may be a human cultural and spiritual reliance of Chinese; romance may be the living style of young generation.

In this case, through his comprehension on the life and human culture, the designer has started his description from the space expansion and taken the dining room as the conjunction point of the space highlights and human active stream lines by the design technique of “sight changing with step moving” to make the visual expansion on the dining room from the space layout. The materials such as glass and dark brown mirror etc. are applied for space partition and the day-lighting requirements of the dining room are also considered.

Romance requires color and fashion atmosphere. Thus comes to the crystal-like transparent glass washrooms, the warm colors, fashionable ornaments and furniture; Lotus- this human cultural subject has run through the entire design of the dining room and is arranged in the middle of the background wall in the dining room. It becomes the most shining color and visual focus in the houses and is performing “very Chinese” romance of lotus.

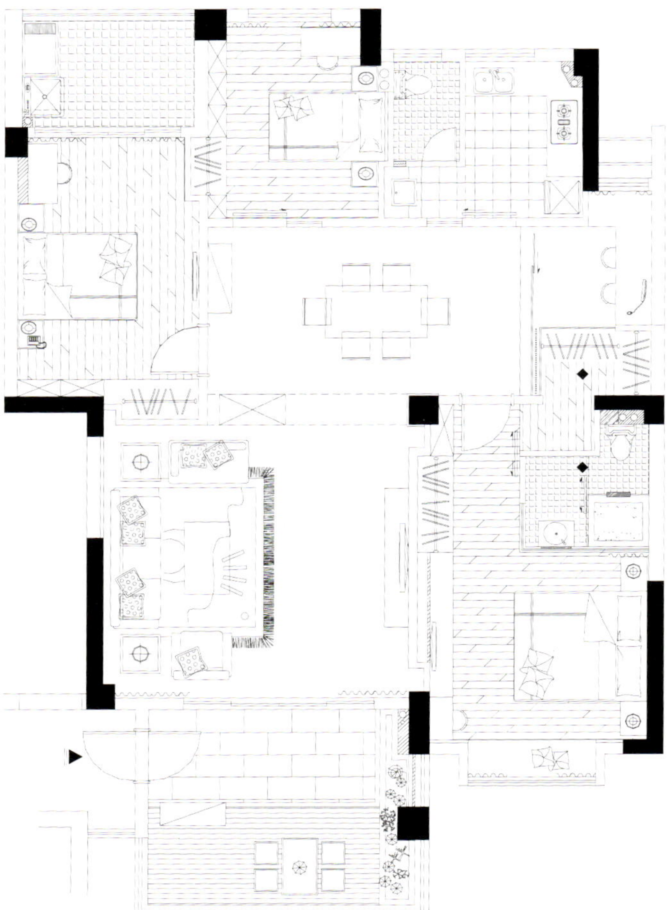

书桌和书架由于使用同种材质，就像一个完整的整体。

The desk and book shelf look integral due to the application of same material.

从客厅看餐厅。

See the dining room from the living room.

湖光山色 A landscape of Lakes and mountains

设计单位:宁波市高得装饰设计有限公司 / 设计:范江 / 参与设计:丁伟哲 / 面积:30 m^2 / 主要材料:彩色贴膜、白色人造石、白色烤漆玻璃、白色防水涂料、塑胶地板 / 坐落地点:宁波 / 工程造价:8万元 / 完工时间:2009年4月

Design Unit:Gotel Design Studio Limited. / Design:Fan jiang / Participatory Design:Ding Weizhe / Area:30 m^2 / Main Materials:color pad pasting、artificial stone、storing varhish glass series、underwater paint、plastic floor / Location:Ningbo / Project Cost:RMB 80 000 / Finish Time:Apr,2009

这是个单身公寓样板房，客户群主要是年轻人，在设计上强调轻松、时尚、活力，在单纯中除了保持卫生间的私密性，其余则还原了所有的东西，衣食住行都一览无余，属完全的自我空间。墙面色彩是不同绿色的渐变，如起伏的青山，如渗透的绿水，更是飞扬的青春，与白色家具的简单恬静构成舒适、温馨，充满生活气息。设计不进行拔高或刻意描绘，但引起客户“入住”的欲望，此楼盘一经推出格外受人欢迎。

It is a show room of an apartment for single men or women. Its customers are mainly the young people and its design emphasizes on relaxation, fashion and vigor. In addition to the reservation of the intimacy of the washroom in the simplicity, the others have restored everything. The basic necessities of life can be taken in everything in a glance and it is a completely individual space. The different green colors are changed gradually on the wall, just like the undulated green mountain and the osmotic green water, which have constituted a comfortable, warm and living atmosphere together with the simplicity and peacefulness of the white furniture. This design does not upgrade or intentionally describe, but has inspired the customers' "staying" desire. This real estate has been extremely popular when it comes into market.

小空间样板，镜子似乎是不可或缺的元素。

In the small show room, mirror seems to be an indispensable element.

进门便一览无余。

Everything can be taken in a glance upon entry.

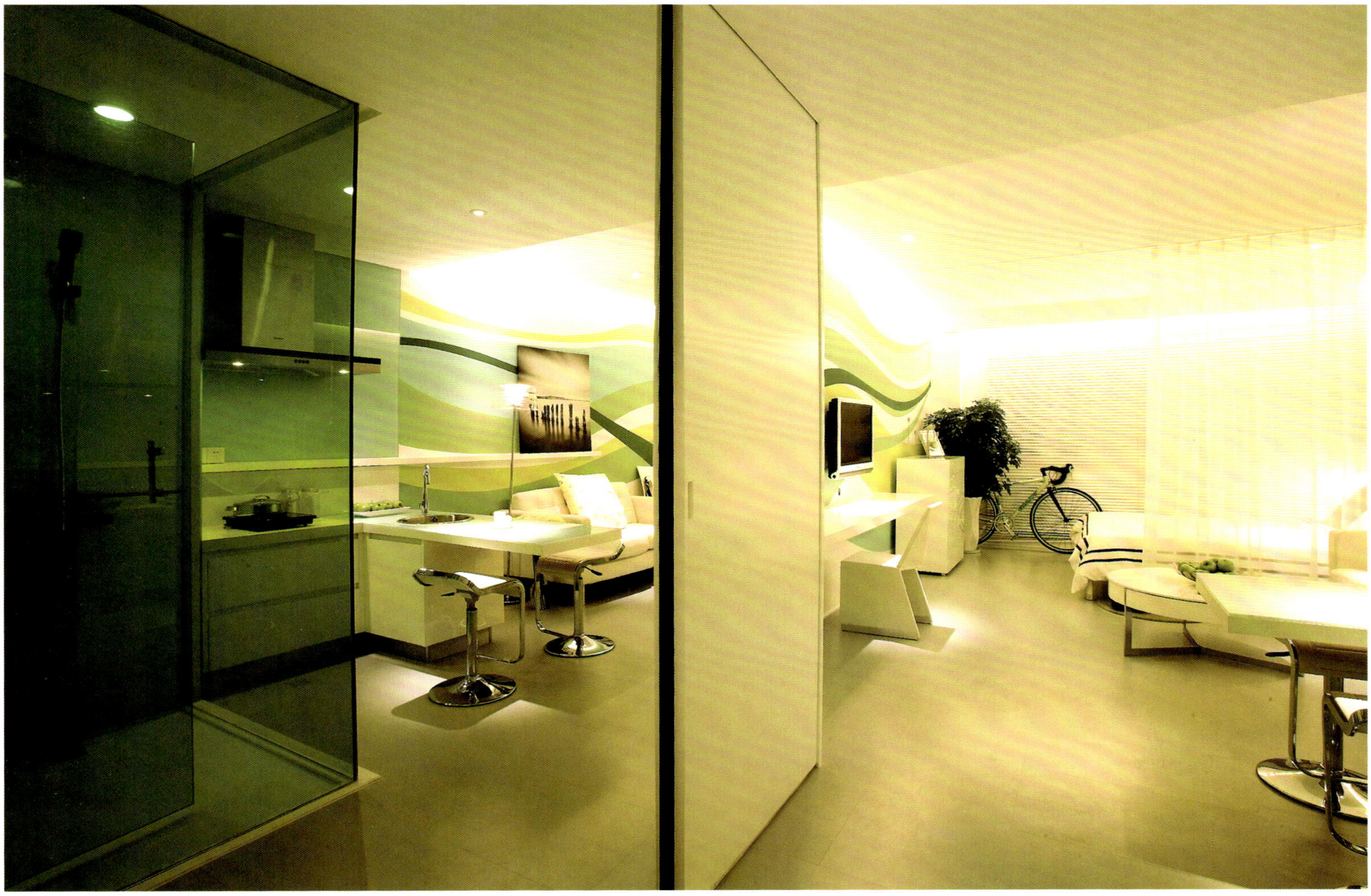

白色顶面周边的茶镜丰富了空间层次。
The brown mirror around the white top has enriched the space levels.
客厅素雅、文静。
The living room looks elegant and quiet.
玻璃移门弱化了功能空间之间的界限。
The sliding glass door has weakened the boundaries among the functional spaces.
简洁、大方的客厅背景墙。
The concise and grand background wall in the living room.
水晶般透明的玻璃卫浴间在主卧房里面。
The crystal transparent glass washroom is located in the master' s bedroom.

不同绿色的渐变如飞扬的青春。

The different green colors are changed just like flying youth.

白色纱帘算是卧室的唯一阻挡。

The white yarn curtain may be the only block in the bedroom.

客厅、厨房、卫生间因狭长的布局而位于同侧。

The living room, kitchen and washroom are located in the same side due to the narrow and long layout.

简单的厨房用具搭配两张椅子，餐桌、操作台都解决了。

The simple kitchen wares in arrangement with two chairs have settled the problems of dining and operation tables.

万科十七英里 THE PENTHOUSE AT 17 MILES

设计单位:香港德坚设计公司 / 设计:陈德坚 / 面积:250 m² / 主要材料:玻璃、木材、天花镜钢、爵士白云石 / 坐落地点:深圳 / 完工时间:2009年3月 / 摄影:陈维忠

Design Unit:Kinney chan and associates / Design:Kinney chan / Area:250 m² / Main Materials:glass、wood、marble / Location:Shenzhen / Finish Time:Mar,2009 / Photography:Chen Weizhong

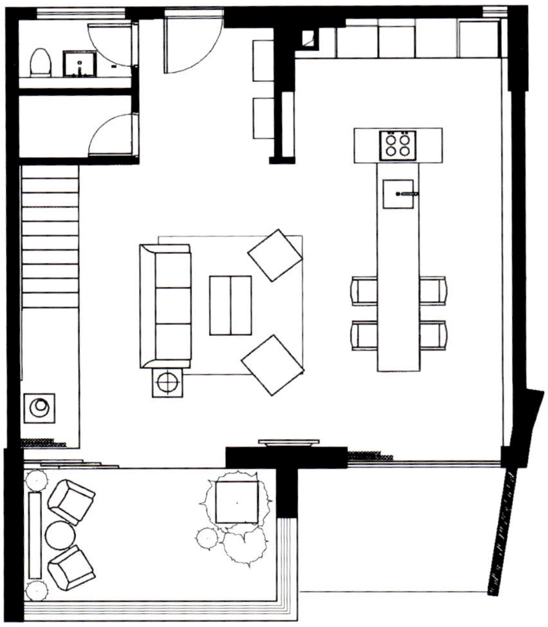

蔚蓝的夜色，蔚蓝的海。

The sky blue night, sky blue sea.

悬浮式木梯从云石墙中伸展出来，配合通透的强化玻璃围栏，形成一组重叠出现的几何线条。

The suspended wood staircase is outspread from the marble wall and forms a group of overlapped geometric lines in combination with the transparent toughened glass railing.

整个空间是面向中国八大最美丽海滩之一的深圳东部大鹏半岛。设计师为了让屋主即使趟在客厅的地毡上，也有被海洋包围的感觉，特地在落地玻璃顶部至天花位置及侧墙加装天花镜钢，加强海蓝色调，为屋主制造最轻松的度假心情。

特色空间由3个楼层打通而成，地下至一楼的悬浮式木梯从云石墙中伸展出来，配合通透的强化玻璃围栏，形成一组重叠出现的几何线条。设于一楼的套房连接露台，床背板以镜面反映空间，床尾则以强化玻璃作间墙物料，既打破以镜子用于睡房的传统，亦可贯通一楼与地下楼层的视野。套厕格局亦不常见，独立分隔的坐厕区及淋浴区，既将浴室功能一分为二，亦将面盆移近房间，配合双面镜子设计，营造不一样的卫浴环境。三楼的主人套房连贯开放式书房设计，以皮制睡床及家具配衬咖啡调的墙砖床背，书房顶部的天窗设计最为特别。这个设于物业最顶层的复式别墅，以天窗引入闪闪发光的星空，好让屋主在假期中，静下来欣赏自然环境，解开身心烦恼。

另外，与主人房左右分隔的套厕连衣帽间，以卡撒灰云石营造豪宅另一焦点。一体化的面盆连浴缸，配合同一物料的墙身及地台设计，视觉上非常简洁。置身于瑰丽的浴室内，无不想来一个悠然的热水浴，再赤脚走出露台，踩在布满小孔的云石地台上，让水柱冲出来按摩脚掌。

Located by the seaside of Shenzhen, PRC, the magnificent sea views were the inspiration behind the interior design scheme of this 3-story residence. The design challenge for this project was to maximize the use of a marvelous sea view feature, and to create comfort that lies the pure simplicity of the residence. The result was a simplistic but spacious design.

The water theme is continued with the use of glass and mirror, making the residence more spacious and natural, with a contemporary feel comes into play in the space.

For the combined living & dining room of this home, designer chooses a charming white marble, used for the flooring through out, forms the basis of an easy, neutral color palette. The marble has been extended up onto the wall where it is teamed with weaved leather paneling, marble stone and exotic timber veneers to create stunning feature wall.

The use of mirror and shiny, reflective surfaces extends the ocean vista into the residence. The focal point of this residence is its windows that look out across the sea. The mirror ceiling of the bedroom adds the illusion of the height to the ceiling and most importantly it reflects the beautiful natural scene into the room.

Simple layout, lighting, light tone color and the use of special shinny materials generate a simple but luxurious atmosphere for the living space.

递进式内部空间布局。

The step-by-step interior space arrangement.

将面盆移近房间，配合双面镜子设计，营造不一样的卫浴环境。

The washbasin is moved close to the room and creates a special washing environment together with the double-faced mirror.

云石墙面经由特殊设计将竹子的造型搬进室内。

The marble wall has moved the bamboo model into the indoor through a special design.

开敞式空间设计。

The open space design.

床背板以镜面反映空间，床尾则以强化玻璃作间墙物料。

The bed backboard reflects the space by mirror while the bed end applies the toughened glass as partition wall material.

蓝调—西城美墅 West Villa (Blues)

设计单位:于强室内设计师事务所 / 设计:于强、林志刚 / 面积:73 m² / 主要材料:蓝色夹胶玻璃、蓝镜、白色烤漆冲孔板、黑色多层实木复合地板、墙纸 / 坐落地点:杭州 / 工程造价:33万元 / 完工时间:2008年11月

Design Unit:YuQiang & Partners Interior Design / Design:Yu qiang、Lin Zhigang / Area: 73 m² / Main Materials:blue laminated glass、blue mirror、white painting perforated plate、black parquet、wallpaper / Location:Hangzhou / Project Cost:RMB 330 000 / Finish Time:Nov,2008

蓝色代表宁静、代表远离尘世的喧嚣。身为现代人的我们，无法逃离都市给我们的生存空间，希望家能成为一个放松心情的宁静港湾。蓝色代表精致、代表品位。我们是年轻的，但我们并不青涩、肤浅、粗糙，我们追求精致的生活，精致是一种品位。蓝色代表思考。我们是年轻的、激情的，更是思考者，经常在灯下探讨人类的智慧。蓝色不是我们唯一的色彩，但这一刻，我们心情是蓝调。

Blue can indicate calmness, and the distance from clamor. Modern people cannot get away from the living space in cities. Therefore, it is hoped that our home can be a quiet place to relax your mind. Blue indicates delicacy and taste. Despite our young age, we are not immature, shallow or careless. We are striving for delicate life. Delicacy represents the taste. Blue indicates the state of thinking. We are young, passionate, but thoughtful, and often discuss the wisdom of human beings late in the evening. Blue is not the only color that distinguishes us; however, we have a mood in blue at the moment.

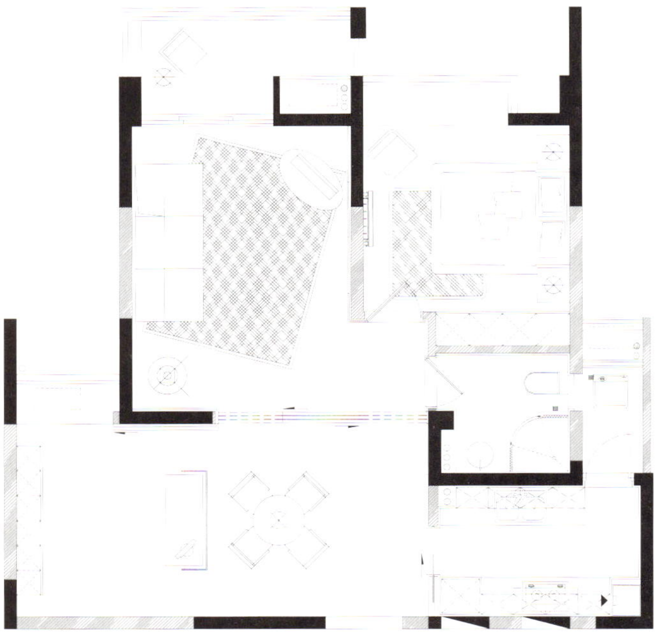

蔚蓝书海中畅游。
Swimming in the blue book sea.
镂空的白色隔断仿佛许许多多思绪的小泡泡。
Hollow and white partition walls like a lot of thinking bubbles.
从顶上挂下来的台灯可以随需要改变角度。
Reading lamp from the ceiling can be adjusted into any angle as required.

华元和庄B1 Hezhuang B1

设计单位:香港方振华设计有限公司、杭州谷谷装饰艺术有限公司 / 设计:方振华 / 参与设计:夏银苹、杨世雄 / 面积:462 m² / 主要材料:大理石、马赛克、墙纸、纱帘 / 坐落地点:杭州 / 工程造价:462万元 / 完工时间:2009年1月

Design Unit:Patrick Fong Design (Hong Kong) Ltd.、Hangzhou GuGu Interior Display Co.,Ltd. / Design: Patrick Fong / Participatory Design:Shirley Xia、Kevin Yang / Area:462 m² / Main Materials:marble、mosaic、wallpaper、sheer curtain / Location:Hangzhou / Project Cost:RMB 4 620 000 / Finish Time: Jan,2009

餐椅造型犹如舞动芭蕾的曼妙女子。
The dining chairs look like a lithe and graceful lady dancing ballet.
银色马赛克跟边的石材地面铺上了新西兰手工羊毛地毯。
The silver mosaic edged stone ground is covered with the handmade wool carpet from New Zealand.

B1户型采用欧式新古典风格。银色马赛克跟边的石材地面铺上新西兰手工羊毛地毯，沙发搭配双层经典欧式绒布窗帘，装饰品将整个奶白色为基调的客厅装点得品位脱俗。餐椅造型犹如舞动芭蕾的曼妙女子，跟圆形餐桌一起在水晶吊灯的映衬下妩媚动人。主卧室构思大胆的床背景，优雅柔媚的床尾凳，亮银色画框里精心绘制的油画跟高档丝绒床品一起构筑奢华雅致的空间。值得一提的是主人房走入式大衣库的设计，超大的衣帽存储空间，可移动的玻璃首饰柜，香槟金装饰的梳妆台，还有三面一体的落地式穿衣镜，想必女主人会因为自己拥有这样的大衣库而雀跃。粉色跟白色构成了女孩房温馨的小天地，抽象暗花图案墙纸搭配水晶壁灯和进口纱帘，体现浪漫情调。

B1 flat-type has applied the continental neoclassic style. The stone material floor with silver mosaic edges is covered with New Zealand handmade wool carpet. The sofa is arranged with the double-layer classic continental-style flannelette window curtain while the furnishings have decorated the whole dining room mainly in creamy white with a refined taste; the model of the dining chairs are just like the pretty women dancing ballet and look charming and attractive together with the round dining table under the setoff of the crystal pendant lamps; in the master's bedroom, the boldly designed bed background, the elegant and lovely bed end bench, the delicately drawn oil picture in the bright silver picture frame as well as the superior velour bed suite have constituted a luxurious and elegant space; what is worth mentioning is the design of a big walkable storeroom in the master's bedroom: the super large clothing and ht storage space, the movable glass ornament cabinet, the champagne dresser. Naturally, the mistress will be very joyful for owning such a big storeroom. Pink and white have constituted a small warm world in the girl's room. The wall paper with the abstract dark flower pattern in combination with the crystal wall-mounted lamps and imported voile curtain has shown a romantic atmosphere.

造型另类的餐椅和衣架。

Dining chairs and clothes rack with special shapes.

卧室需要一份神秘和宁静。

The bedroom also needs some mystery and quietness.

茶几也可以坐在地上。

The tea table can also sit on the floor.

温馨、浪漫的卧室。

The warm and romantic bedroom.

走入式大衣库，想必女主人会因为自己拥有这样的大衣库而雀跃。

For the big walkable clothing storeroom, the mistress will naturally be very joyful for owning such a big storeroom.

书房十分注意室外光线的引入。

The study room has paid great attention to the introduction of outdoor light.

主卧室床背景构思大胆。

The bed background in the master’s bedroom is designed boldly.

荣泰河庭 Rongtai Heting

设计单位:RYB・三原色建筑装饰设计院 / 设计:符军 / 参与设计:唐锦同 / 面积:150 m² / 主要材料:清玻、马赛克、艺术花砖、珠帘、墙纸、实木地板 / 坐落地点:珠海 / 工程造价:33万元 / 完工时间:2008年10月 / 摄影:周星

Design Unit:RYB architectural design & decoration institute / Design:Fu jun / Participatory Design:Tang Jintong / Area:150 m² / Main Materials:mosaic、brick、wallpaper、solid wood flooring / Location:Zhuhai / Project Cost:RMB 330 000 / Finish Time:Oct,2008 / Photography:Zhou xing

设计的挑战在于完善房屋的实用功能并营造非同寻常的开阔体验，借此更好地展示内建筑的固有元素。客厅与书房间的玻璃、珠帘，主卫与主卧间的玻璃、珠帘，使得整个空间的比例更为和谐。过道墙面、地面安排艺术花砖，赋予了空间更多的变化。

The challenge of the design lies to improve the practical functions of the residence and create an unusual wide experience in order to better display the inherent elements inside the building. The glass and bead curtain between the living room and study room as well as between the master' s room and its washroom has have made the proportion of the whole space more harmonious. The wall and floor of the passageway are covered with artistic patterned tiles to endow the space with more changes.

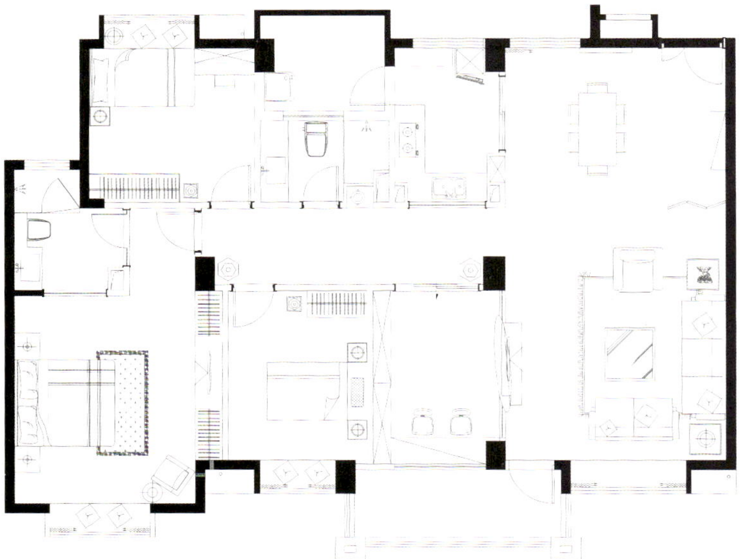

精雕细琢的电视背景墙。
The delicately and finely arranged TV background wall.
纯白的空间点缀些许黑色家具、配饰。
The pure space is dotted with some black furniture and ornaments.

山水云房LOFT Fortune Courtyard LOFT

设计单位:上海汉星设计+熠彩软装 / 设计:高龙 / 面积:78 m² / 主要材料:地板、大花白大理石 / 坐落地点:南京 工程造价:25万元 / 完工时间:2009年6月 / 摄影:高龙

Design Unit:Shanghai HaiXing Design+YiCai Soft Decoration / Design:Gao long / Area:78 m² / Main Materials:floor、marble / Location:Nanjing / Project Cost: RMB 250 000 / Finish Time:Jun,2009 / Photography:Gao long

在设计时利用房型特有的挑高，将书橱做成入墙式，形成顶天入地的感觉，给人带来强烈的冲击力，形成视觉中心点。

地板做旧，体味自然的同时，又融入些许生活的元素。遵循简色主义，除了黑、白、原木这三种颜色，空间里几乎看不到多余颜色。餐桌椅、工作台都选用原木色，连地毯也采用低调的黑白条纹。此外，梯与镜的结合为空间增色不少。

该案点睛之处在于半开放式卧室的处理。床被抬高，做成榻榻米，设计师将床背对客厅，在某种程度保持了一定的私密性。也正因为半开放，设计师还巧妙地将床头背后的空间处理成一个简单的吧台，夜深人静，工作累了还可以小饮一杯，转过身来即可入睡。

In the design, the unique overhanging flat type is applied to embed the book shelf into the wall and form a feeling of standing on the ground with head up to the sky, which has brought the strong impact on the people and becomes the visual center.

The floor boards are made antique, which have been integrated with certain living elements while enjoying the nature. It has followed the principle of simple colors. You can hardly see redundant colors in the space except black, white and original wood. The dinging table, chairs and working table have been selected with original wood color and even the carpet has applied to the low-pitched black and white strips. Furthermore, the combination between the staircase and mirror has given an added grace to the space greatly.

The highlight of this case is the arrangement of the semi-opening bedroom. The bed is uplifted and made into a tatami. The designer puts the bed against the living room to keep the certain intimacy to some degree. Just because of its being semi-opening, the designer also tactfully makes the space behind the headboard into a simple bar counter. In the still of the night, you can take a drink and then turn around to go to sleep when working tired.

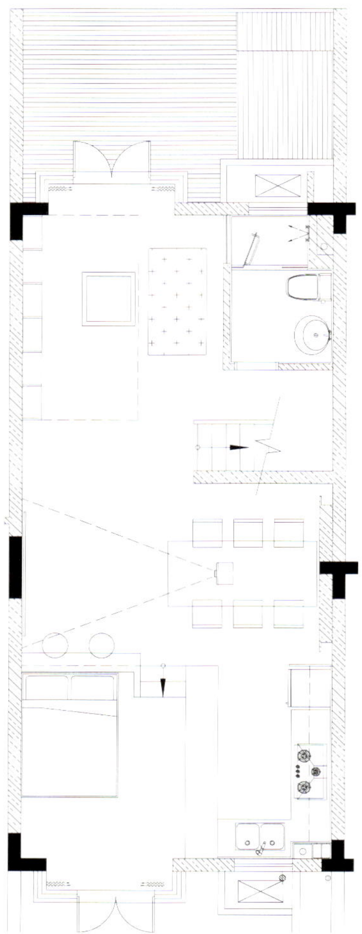

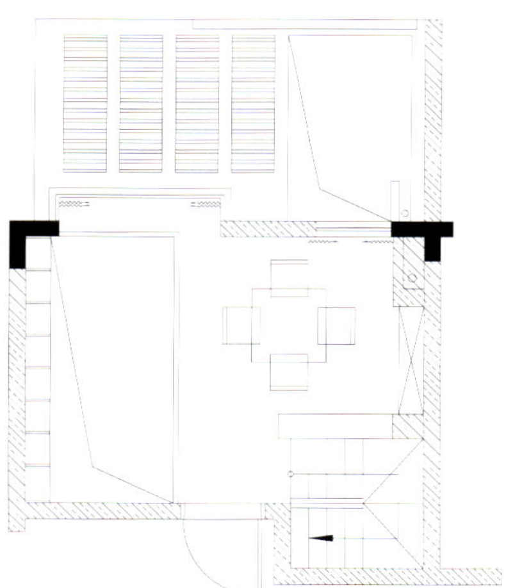

特有的房型挑高。

The unique overhanging flat-type.

将书橱做成入墙式，形成顶天入地的感觉。

The book cabinet is made in a wall type to form a feeling of being top to the sky and bottom to the ground.

CCTV 6

清玻璃的置入扩充了空间。
The arrangement of clear glass has expanded the space.
简洁的客厅。
The concise living room.
深浅两种色调搭配的过道。
The corridor with the proper combination of dark and light colors.

山水云房LOFT Fortune Courtyard LOFT

设计单位:上海汉星设计+熠彩软装 / 设计:高龙 / 面积:78 m² / 主要材料:地板、大花白大理石 / 坐落地点:南京 工程造价:25万元 / 完工时间:2009年6月 / 摄影:高龙

Design Unit:Shanghai HaiXing Design+YiCai Soft Decoration / Design:Gao long / Area:78 m² / Main Materials:floor、marble / Location:Nanjing / Project Cost: RMB 250 000 / Finish Time:Jun,2009 / Photography:Gao long

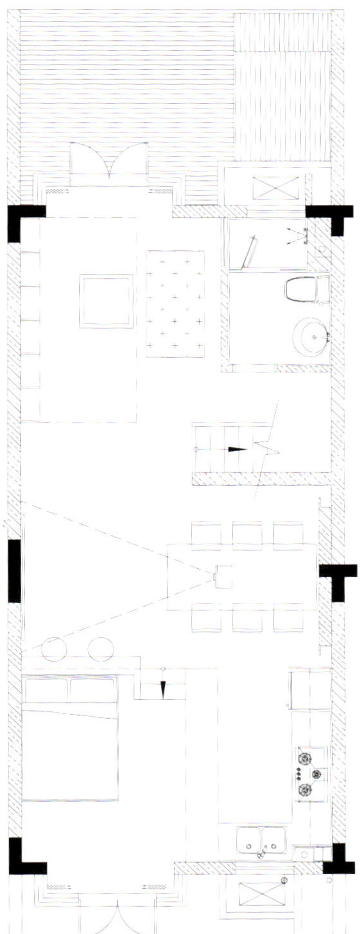

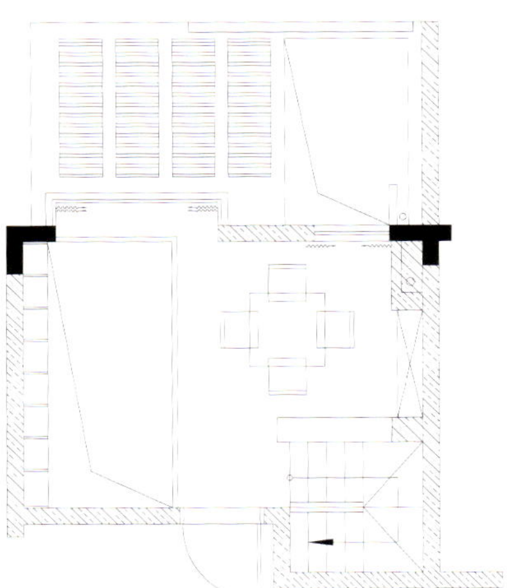

特有的房型挑高。

The unique overhanging flat-type.

将书橱做成入墙式，形成顶天入地的感觉。

The book cabinet is made in a wall type to form a feeling of being top to the sky and bottom to the ground.

在设计时利用房型特有的挑高，将书橱做成入墙式，形成顶天入地的感觉，给人带来强烈的冲击力，形成视觉中心点。

地板做旧，体味自然的同时，又融入些许生活的元素。遵循简色主义，除了黑、白、原木这三种颜色，空间里几乎看不到多余颜色。餐桌椅、工作台都选用原木色，连地毯也采用低调的黑白条纹。此外，梯与镜的结合为空间增色不少。

该案点睛之处在于半开放式卧室的处理。床被抬高，做成榻榻米，设计师将床背对客厅，在某种程度保持了一定的私密性。也正因为半开放，设计师还巧妙地将床头背后的空间处理成一个简单的吧台，夜深人静，工作累了还可以小饮一杯，转过身来即可入睡。

In the design, the unique overhanging flat type is applied to embed the book shelf into the wall and form a feeling of standing on the ground with head up to the sky, which has brought the strong impact on the people and becomes the visual center.

The floor boards are made antique, which have been integrated with certain living elements while enjoying the nature. It has followed the principle of simple colors. You can hardly see redundant colors in the space except black, white and original wood. The dinging table, chairs and working table have been selected with origina wood color and even the carpet has applied to the low-pitched black and white strips. Furthermore, the combination between the staircase and mirror has given an added grace to the space greatly.

The highlight of this case is the arrangement of the semi-opening bedroom. The bed is uplifted and made into a tatami. The designer puts the bed against the living room to keep the certain intimacy to some degree. Just because of its being semi-opening, the designer also tactfully makes the space behind the headboard into a simple bar counter. In the still of the night, you can take a drink and then turn around to go to sleep when working tired.

Arletta

地板做旧，体味自然的同时，又融入些许生活的元素。

The floor boards are made antique to be integrated with certain living elements while experiencing the nature.

床被抬高，做成榻榻米。

The bed is raised up to form a tatami.

地毯采用低调的黑白条纹。

The carpet has applied the unobvious black and white stripes.

遵循简色主义，空间里几乎看不到多余颜色。

The color simplism is followed and it is hardly to seen any redundant color in the space.

哈公馆 Harbin Residence

设计单位:哈尔滨唯美源装饰设计有限公司 / 设计:靳全勇 / 面积:280 m^2 / 主要材料:大花白石材、皮革、焗漆玻璃、壁纸 / 坐落地点:哈尔滨 / 工程造价:280万元 / 完工时间:2009年3月 / 摄影:张奇永

Design Unit:WeiMeiYuan Design / Design:Jin Quanyong / Area:280 m^2 / Main Materials:stone、leather、coated glass、wallpaper、leather / Location:Ha Erbin / Project Cost:RMB 2 800 000 / Finish Time: Mar,2009 / Photography:Zhang Qiyong

整体空间以块和面的形式划分区域，以黑白色为基调，配合暖灰色墙面，使其在现代简约的基础上又多了些优雅与华贵。陈设大量采用不锈钢材质替代传统的材质，给人耳目一新的感觉。地道的形体结构、简单的线条与设计风格融为一体，彰显时尚前卫的生活方式，让业主在每个角落里都能产生欣赏与享受的愉悦。

The whole space is divided by block and face uses black and white as main colors. Together with the warm gray wall, it is also endowed with some elegance and luxury in addition to the modernity and simplicity. Stainless steel materials are used to replace the traditional materials greatly for the displays to provide a refreshing feeling. The pure physical structure and simple lines are integrated with the design style to prominently deliver a fashionable and modern living style so that the owners can have an enjoying and entertaining pleasure at every corner.

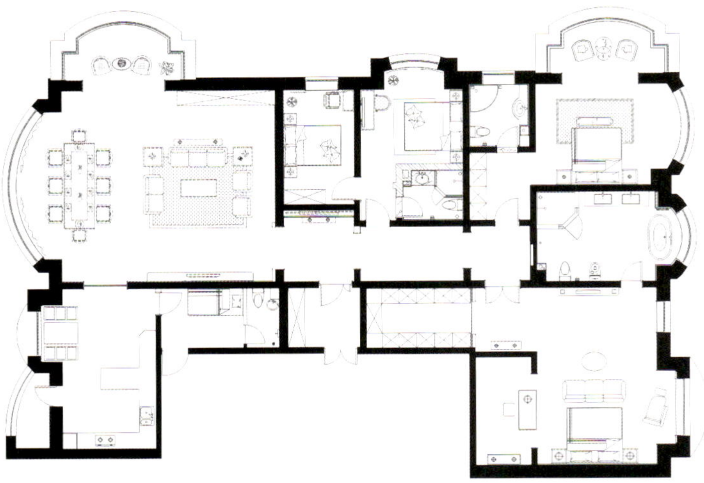

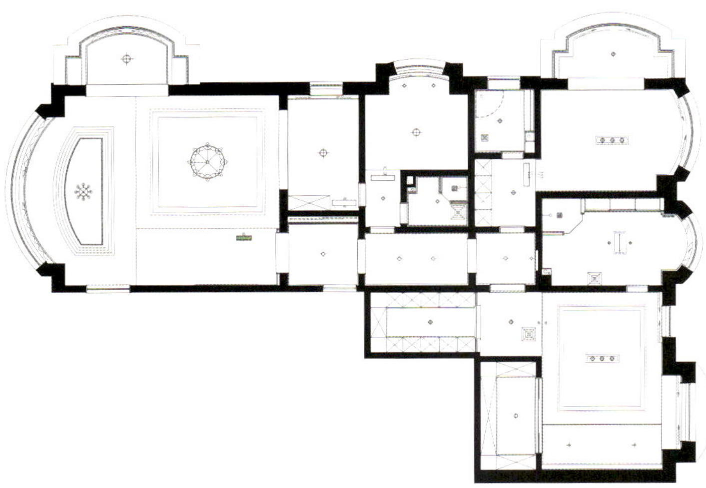

采用不锈钢材质替代传统的材质。
The stainless steel material is applied to replace the traditional material.
地道的形体结构彰显时尚前卫的生活方式。
The genuine structure has prominently shown the fashionable and pioneering living style.

大花白石材连接了地面和顶面。
The large gray stone materials have connected the ground and ceiling.
深棕色的卧室里，灯光打头阵。
In the dark brown bedroom, the light takes the lead.
卫生间被大花白石材包围。
The washroom is surrounded by the large gray stone materials.
舍弃大吊灯的做法满有意思。
The way of giving up the large pendant lamp is quite interesting.

简单的线条与设计风格融为一体。
The simple lines are integrated with the design style.
光影的魅力赋予空间华贵的气息。
The charm of light and shadow has endowed the space with a luxurious atmosphere.
三个水晶吊灯排齐而挂，夸张。
Three crystal pendant lamps are hung in a row, exaggerating.

玩味视觉空间 Rumination on Visual Space

设计单位:高文安设计有限公司 / 设计:高文安 / 面积:380 m² / 主要材料:云石、胡桃木 / 坐落地点:香港 / 完工时间:2008年9月

Design Unit: KENNETH KO COMPANY / Design:Kenneth Ko / Area:380 m² / Main Materials:marble、walnut wood / Location:Hong Kong / Finish Time:Sep,2008

这个三层复式的单位采用现代古典的风格，当别人将焦点集中于空间装饰，高先生则把着重点放在建筑架构之上，让空间呈现不凡美感，犹如玄关楼梯设计把三层的空间连成一体，令进门口的视野向上延展，让访客的视线自然地向上转移，缔造出层层递进的视觉趣味。

玄关右侧便可看到餐厅和客厅，采用米、白色作为主调，更在餐厅的正面墙利用米白色的云石，显得高贵大方。

在单调的墙纸上托印出古典的图案，更在客厅的位置设计一个线条简单而高贵的壁炉，配上客厅的水晶灯表现出古典的韵味，位处一层更有一家庭厅，利用原木色的木饰面，带出柔和的气氛。

一层还有两个孩子房、客房、浴室和工人房，在顶层是全然打通式规划的主卧，划分为寝区、工作区、浴室和衣帽间四个部分。工作区更利用一大型的镜门把书房和寝区分隔，面向主卧与客厅景观开扬，尽展超然气派。

This three-story compound type house has applied the modern classic style with the emphasis on its building structure to display the unusual beauty of the space. The staircase at the front door is designed to connect the three spaces into a whole so that that visual field of the entrance extends upward to guide the eyes of the visitors to move upward naturally, which creates a layer-by-layer visual pleasure.

The dining room and living room are located on the right side of the front door with beige and white as main colors, which looks noble and grand.

The wall paper is printed with convex patterns. Moreover, a fireplace is designed in the living room in an arrangement with a crystal lamp to show a classic charm. The sitting room has applied the original wood-colored veneers to generate a tender atmosphere.

There are also two rooms for children, guest room and bathroom as well as maid room at the first floor. At the top floor is the master' s bedroom that is completely open including the sleeping area, working area, bathroom and cloakroom. The working area has been applied with a large mirror door to separate the study room and sleeping area. The scenes of the master' s bedroom and living room are quite opening with a full revelation of the transcendent style.

举目远眺，美景尽收眼底。
Overlooking faraway, you can have a panoramic view on the beautiful scenery.
落地玻璃很有风度地让室外美景融入室内。
The floor glass gracefully integrates the outdoor beautiful scenery into the indoor.
客厅的水晶灯表现出古典的韵味。
The crystal light in the living room delivers a classic charm.

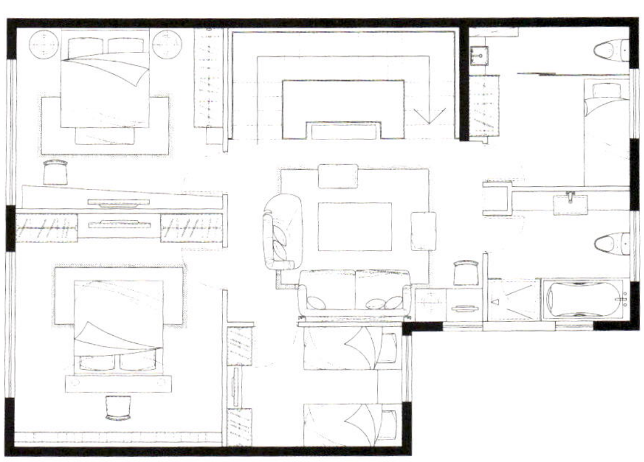

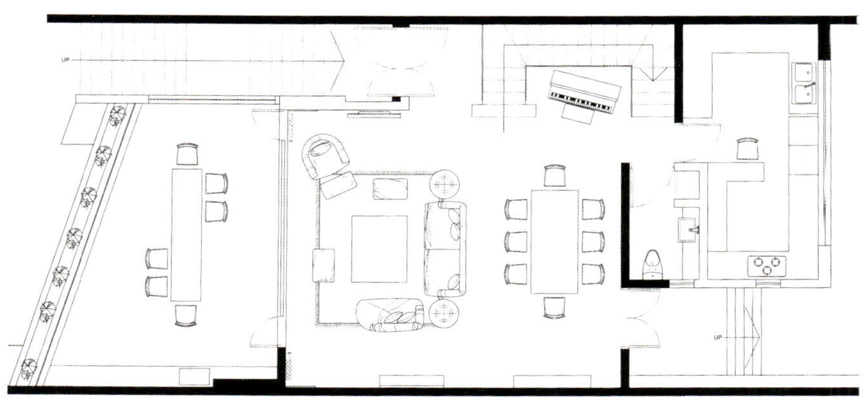

家庭厅采用米、白色作为主调。
The family hall has taken off-white and white as its main colors.
利用原木色的木饰面，带出柔和的气氛。
The original wood veneers are applied to bring out a gentle atmosphere.
利用镜当门，把书房和寝区分隔。
The mirror is used as a door to separate the study room from the sleeping area.
卫浴间也能看电视。
You can also watch TV programs in the washroom.
卧室通畅、整洁。
The bedroom is clear and clean.

清合嘉源 Qing He Jia Yuan

设计单位:杭州观堂室内设计有限公司 / 设计:张健 / 参与设计:林子全 / 面积:124 m² / 主要材料:砖、木板、墙纸、马赛克、强化地板 / 坐落地点:杭州 / 完工时间:2008年11月 / 摄影:王飞

Design Unit:Guan Interior Design Co.,Ltd. / Design:Zhang jian / Participatory Design:Lin Ziquan / Area:124 m² Main Materials:brick、plank、wallpaper、mosaic、laminated flooring / Location:Hangzhou / Finish Time:Nov,2008 / Photography:Wang fei

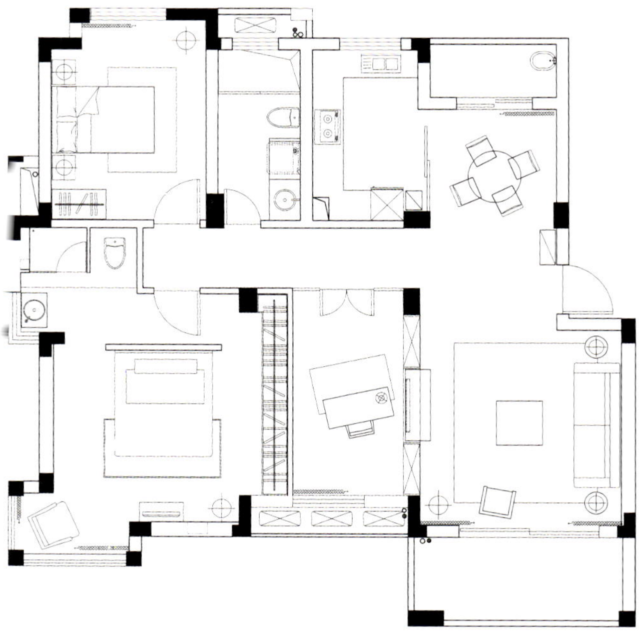

洁净、纯粹的过道。
The clean and pure corridor.
圆形餐桌配以白色餐椅，突出精致的生活气质。
The round dining table is arranged with the white dining chairs to stress the delicate living atmosphere.
马赛克贴面茶几为专门定制，彰显品位。
The mosaic veneered tea table is specially customized and conspicuously shows the taste.

本案为三房两厅两卫格局。风格定位在简欧、偏现代，具有Art Deco风格。
客厅背景墙整面采用砖墙材质，与简欧的整体风格形成混搭；马赛克贴面茶几为专门定制，彰显品位。结合餐厅空间有限的特点，在软装设计时，特意选用圆形餐桌，配以白色餐椅，突出精致的生活气质。客厅与书房之间的墙体采用半透明隔断，部分为实体方格，可以摆放书刊与饰品，部分采取透明玻璃，将空间连通。主卧入口处一道隔墙，起到玄关作用，为主人提供更隐秘的空间。墙纸延续书房与过道的整体风格，淡雅大方。左侧为衣柜，解决储藏纳物功能；衣柜的四扇超大移门外表铺以正面落地镜，突显大气与整洁。儿童房采用粉色为主。在墙纸、地毯、家具、床品的选择上紧紧围绕这一中心。卫生间采用玻璃马赛克贴面，清爽整洁。

It is a pattern of three bedrooms, two sitting rooms and two washrooms. Its style is orientated as simple continental and quite modern. It has an Art Deco style.
The whole background wall in the dining room has applied the brick wall to arrange with the entire simple continental style; the mosaic covered tea table is specially customized with a prominent display of the taste. The wall between the dining room and study room has applied the translucent partition, part of which is solid grid for placement of the books, magazines and furnishings and part of which is transparent glass to connect the space. There is a partition wall at the entrance of the master's bedroom as the front door in order to provide a more intimate space for the master. The wall paper has continued the entire style of the study room and corridor, simple, elegant and grand. The wardrobe is on the left to settle the storage function; four leaves of super large sliding doors can hide the various types of articles in the wardrobe and are covered with the floor mirror in the front, which look grand and neat. The children's room has applied pink as the main color and the selection of the wall paper, carpet, furniture and bed suite has been closely centered on this. The washroom has applied the glass mosaic facing, fresh and tidy.

主卧入口处一道隔墙，起到玄关作用。

One partition wall at the entrance to the master' s bedroom functions as the front door.

地毯延续书房的整体风格，淡雅大方。

The carpet has continued the whole style of the study room, quietly elegant and natural.

衣柜外表辅以正面落地镜，突显大气与整洁。

The chest is covered with a front floor mirror with prominent display of the greatness and neatness.

朗园样板房 Show Room of Sunny Garden

设计:林开新 / 面积:80 m² / 主要材料:水曲柳染白、白色大理石、玻璃、墙纸、不锈钢 / 坐落地点:福州 / 工程造价:15万元 / 完工时间:2008年11月 / 摄影:吴永长

Design:Lin Kaixin / Area:80 m² / Main Materials:fraxinus mandshurica、marble、glass、wallpaper、stainless steel / Location:Fuzhou / Project Cost:RMB 15 000 / Finish Time:Nov,2008 / Photography:Wu Yongchang

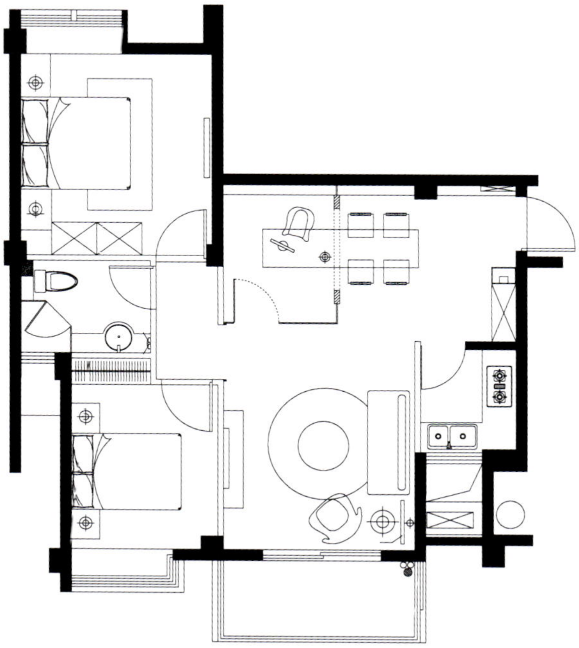

本案定位为现代、简约又富有活力的时尚风格。设计师融汇镜面、英文图式、红白对比色、经典造型家具等元素，使空间无处不流露出鲜明的时尚感觉。

以大面积的白、奶白、灰白为底色，再点缀亮红色的家具、玻璃隔断等，搭配手法大胆而又恰到好处。以墙纸、床品等体现的英文图式进一步充实了内涵和质感；餐厅及次卧室背景墙均选用英文凹凸墙纸，色调单纯而具质感；主次卧也相应选用英文图案的床品，让丰富的色彩跳跃其间。

利用镜面和红色半透明玻璃隔断，在餐厅旁构造出书房，增加实用区域的同时避免了新增区隔给空间带来的局促。在分隔餐厅与书房的镜面下留下长条形的镂空，营造通而有别的空间序列。

This design is positioned in a modern, simple and vigorous fashion style. The designer has collected the elements such as mirror, English figures, the contrasting colors of red and white as well as the classic modeled furniture so that the space has shown a vivid fashion feeling everywhere.

White, creamy white and grayish white over large areas are applied as the basic color and then the bright red furniture and the glass partitions etc. are dotted, which are arranged bold but appropriate. The English figures reflected by the wall paper and bed suite have further enriched the meaning and quality; the background walls in the dining room and secondary bedroom have applied English concave and convex wall papers, which have simple colors and quality senses; the appropriate bed suites with English figures in the master's and secondary bedrooms are also selected to make the rich colors jumping among them.

The mirror and red translucent glass partition are applied to create a study room beside the dining room, which has avoided the constraint brought to the space by the newly added area while increasing the practical area. A long hollow-out is reserved below the mirror separating the dining room from study room to create a connecting but distinguished space sequence.

精致的餐具和配饰。

The delicate table wares and ornaments.

利用镜面和红色半透明玻璃隔断，在餐厅旁构造出书房。

The mirror and red translucent glass partitions are applied to create a study room beside the dining room.

镜面的发射扩充了空间。
The reflection of the mirror has expanded the space.
大面积的白、奶白、灰白为底色，点缀亮红色的家具。
White, creamy white and grayish white in large areas are applied as the basic color to decorate the bright red furniture.
色调单纯而具质感。
The colors are pure and of quality.

广丰新东街-G2户型 Guangfeng Xindong Street — G2 Flat Type

设计单位:江西联动建筑装饰设计工程有限公司 / 设计:童武民、李信伟 / 面积:130 m² / 主要材料:黄洞石、金属丝墙纸、黑木纹大理石 / 坐落地点:上饶 / 工程造价:25万元 / 完工时间:2008年10月 / 摄影:邓金泉

Design Unit:The Design Institute Of Enviromental Art Of Jiangxi Lian Dong / Design:Tong Wumin、Li Xinwei / Area:130 m² / Main Materials:yellow hole stone、wallpaper、marble / Location:Shangrao / Project Cost:RMB 250 000 / Finish Time:Oct,2008 / Photography:Deng Jinquan

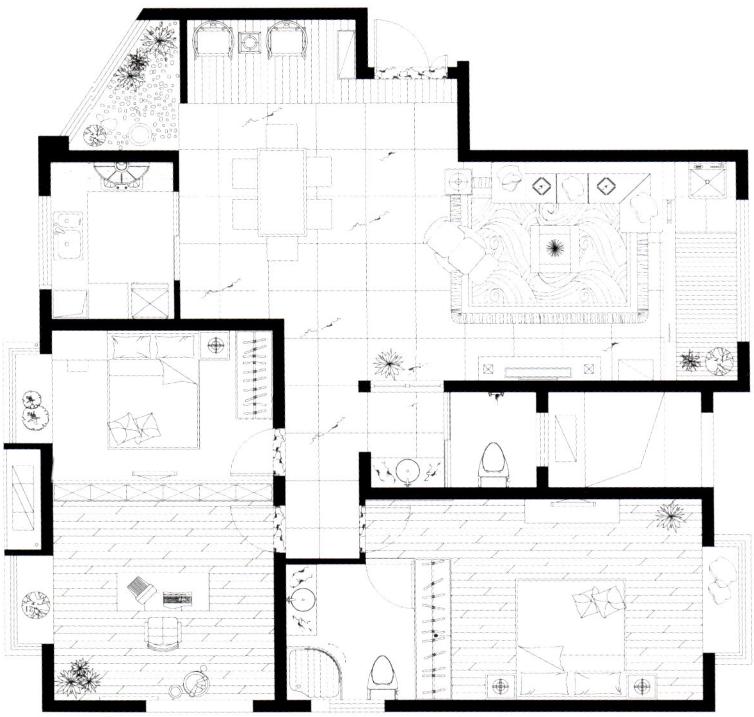

本案无论是室内家具的风格，还是内部装饰，中式传统文化的内涵都被诠释得淋漓尽致。总体布局对称均衡，在装饰细节上富于变化。整体风格上把握大气和稳重的家居氛围，大量使用实木，奠定了中式基调，柔和的质地、淡淡的木香和细腻的纹理，让家中充满了书香门第的雅致。而凳椅、屏风、茶几、书桌、画案这些中式风格元素的融入，彰显出内敛而高贵的风范。优雅的青花瓷恰到好处地点缀了家中各个角落，营造出儒雅的中式文化氛围，为生活增添了一丝淡雅的怀旧感，一种说不出道不明的灵性生活味道。

No matter the style of the interior furniture or the interior decoration, this design has incisively and vividly interpreted the meanings of Chinese-style traditional culture. The general layout is symmetric and balanced and rich in the decorative details. Its general style has grasped the grand and steady family atmosphere and a large quantity of solid woods has established the Chinese- style mood. The soft quality and faint wood fragrance as well as exquisite texture have filled the home with the scholar' s elegance. The application of these Chinese-style elements such as the stools, chairs, screens, tea table, desk and drawing table has prominently shown an introversive and noble demeanor. The elegant blue and white porcelains have just right decorated every corner in the home and created a genteel Chinese-style cultural atmosphere, which has added a faint reminiscence to the life and an unspeakable intelligent life smell.

内部装饰和家具将中式传统文化的内涵诠释得淋漓尽致。

The interior decoration and furniture have incisively and vividly interpreted the meanings of Chinese-style traditional culture.

广丰新东街– G6户型 Guangfeng Xindong Street — G6 Flat Type

设计单位:江西联动建筑装饰设计工程有限公司 / 设计:童武民、李信伟 / 面积:180m² / 主要材料:墙纸、阿曼米黄、爵士白大理石 / 坐落地点:上饶 / 工程造价:48万元 / 完工时间2008年10月 / 摄影:邓金泉

Design Unit:The Design Institute Of Enviromental Art Of Jiangxi Lian Dong / Design:Tong Wumin、Li Xinwei / Area:180m² / Main Materials:wallpaper、amasa beige、marble / Location:Shangrao / Project Cost:RMB 480 000 / Finish Time:Oct,2008 / Photography:Deng Jinquan

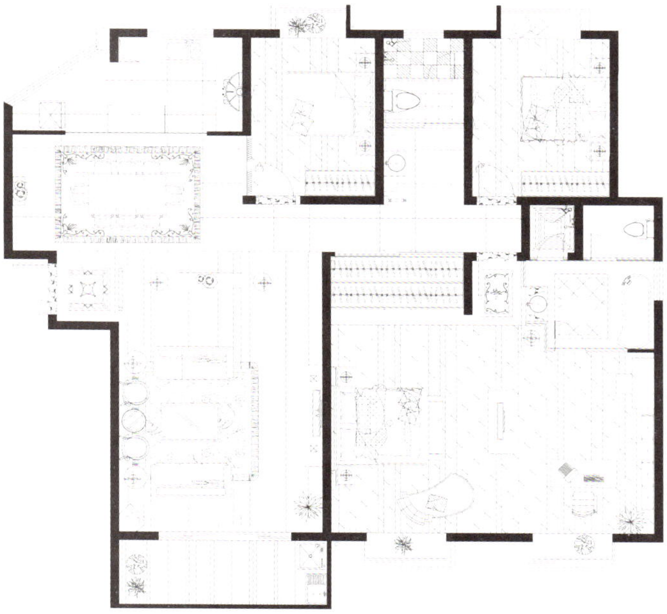

在欧式后现代室内装饰中，设计师最大限度地强调了色感与质感的对比，天然进口石材、进口德国壁纸、工艺玻璃、现代造型的电脑雕饰与舒适功能的完美结合，反映出现代都市人进入后现代社会的另一种思考方向。

In the continental-style postmodern interior decoration, the designer has emphasized the contrast between the color and quality to the maximum. The perfect combination of natural imported stone materials, imported German wall papers, artistic glass, and computerized carving ornaments in modern models with their comfortable functions has reflected another aspect for consideration after the modern urbanites enter the postmodern society.

开放式厨房连接了餐厅。
The open kitchen is connected with the dining room.
精致的细部。
The delicate details.

客厅、餐厅、厨房全局观。
The full view of the living room, dining room and kitchen.
地面在不知不觉中区别开两个功能空间。
The ground has unconsciously divided two functional spaces.
爵士白大理石主打的过道。
The corridor mainly with the jazzy white marbles.
书房和卧室用珠帘相隔。
The bead curtain is applied to separate the study room from the bedroom.

有人喜欢蓝 Someone likes blue

设计单位:明威建筑设计工程（香港）有限公司 / 设计:谢智明 / 参与设计:霍律鸣 / 面积:500 m² / 主要材料:墙纸、地砖、玻璃马赛克 / 坐落地点:佛山 / 工程造价:80万元 / 完工时间:2009年5月 / 摄影:钱翔

Design Unit:MingWay Building Design Engineering (Hong Kong) Limited / Design:Xie Zhiming / Participatory Design:Huo Lvming / Area:500 m² / Main Materials:wallpaper、floor tile、glass mosaic / Location:Foshan / Project Cost:RMB 800 000 / Finish Time:May,2009 / Photography:Qian xiang

在现代城市家居设计里，充分体现业主以家居生活为核心，必然考虑两个重要因素。第一，环境空间，生活其中的人与空间尺度把握合理。第二，人文元素，使业主的爱好与个性融入其中。本来是一家四口的家，首层从功能酒吧、茶区、影视到就餐都合理安排融合，体现夫妇的生活爱好。一幅西班牙图腾马赛克拼图彰显业主的艺术喜好。二层夫妇和子女不同个性风格的房间加上简洁的起居室构成一个温馨的生活区，这就是热爱的家居生活。

The first floor is arranged properly and integrated with a functional bar, tea area, movie and television as well as dining, which have reflected the living favors. Furthermore, the mosaic picture in Spanish totem has also shown the artistic favor of the owner. The individualized rooms for the couple and the children together with the simple sitting room have constituted a warm living area. It is the love of home life.

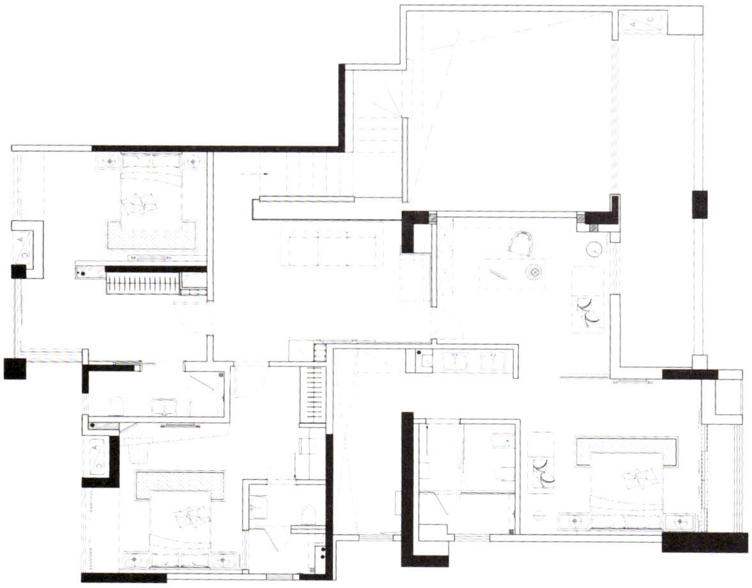

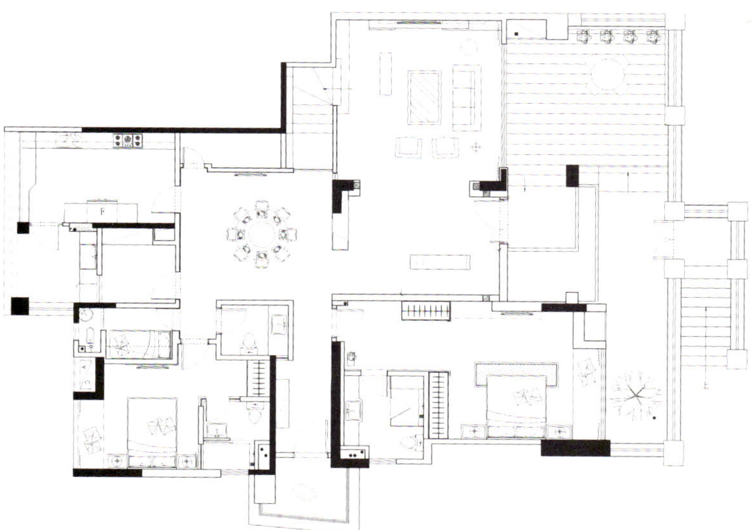

庄重、低调的楼梯拐角。

The grand and low-profiled staircase corner.

大幅落地玻璃引入更多室外光线。

The large floor glass has introduced more outdoor light in.

不被人打扰的视听区。

An audiovisual area without interruption.

一面玻璃马赛克、一面墙纸，繁复与单纯的对比。
One side of glass mosaic, one side of wall paper, the contrast between complication and simplicity.
简单而不失色彩变化的儿童房。
The simple but colorful child's room.
狭长的盥洗区。
The narrow and long washroom.
色调简单的卧室。
The simple colored bedroom.

天俊蓝水湾 Tianjun Blue Bay

设计单位:福州品川设计顾问有限公司 / 面积:135 m^2 / 主要材料:水曲柳面板、艺术墙纸、玻化砖、仿古砖、艺术挂帘、不锈钢、钢化玻璃 / 坐落地点:福州 / 工程造价:12万元 / 完工时间:2009年4月 / 摄影:吴永长

Design Unit:Pure Charm Design Consultants Limited / Area:135 m^2 / Main Materials:fraxinus mandshurica、art wallpaper、glazed tiles、rustic tiles、stainless steel、tempered glass / Location:Fuzhou / Project Cost:RMB 120 000 / Finish Time:Apr,2009 / Photography:Wu Yongchang

作为简约欧式设计作品，在客厅色彩上，选择黑白灰作为主色调。白色地板与黑色墙面形成对比，灰色棉麻沙发则以随性的姿态呈现其中。墙面上埋藏着不规则的内置线性橘色光源，散发着灵动的光彩；电视柜上抽象的树木造型装饰品则以黑色呼应空间，以红色为之增色。

三件在材质和造型上大相径庭的座椅摆放在客厅中，给人带来和谐的质感；灰色棉麻沙发搭配彩色抱枕，无限柔软舒适；墙角白色小桌上置放的绿色台灯，混杂着古典和高科技的美感。卧室设计上注重柔软与温情的感受，黄色作为主色调，尽显柔和。落地灯以简洁的造型散发美感。床边的台灯则以更为柔和的姿态呈现，花纹线条在灯光照耀下更显朴雅的美感。

As a works in a design of simple continental style, black, white and gray are selected as the main colors for the dining room. The white floor and black wall has formed a contrast while the gray cotton linen sofa is placed casually among it. The irregularly embedded linear orange light source is buried in the wall and sends out the bright gloss; the abstract wood sculpture on TV cabinet echoes with the space in black and brightens it in red.

The three sets of chairs widely different in material and sculpture are placed in the dining room and bring people a harmonious atmosphere; the gray cotton linen sofa in arrangement with the colorful cushions looks infinitely soft and comfortable; a green table lamp is placed on the small white table in the corner, mingled with the classic and hi-tech aesthetics. The design of the bedroom has emphasized on the soft and warm feelings. Yellow, as a main color, has shown the full gentleness. The floor lamp has emitted the aesthetics in a simple sculpture. The table lamp beside the bed looks even more softer while the lines of the decorative patterns show a more simple and elegant aesthetic under the illumination of the light.

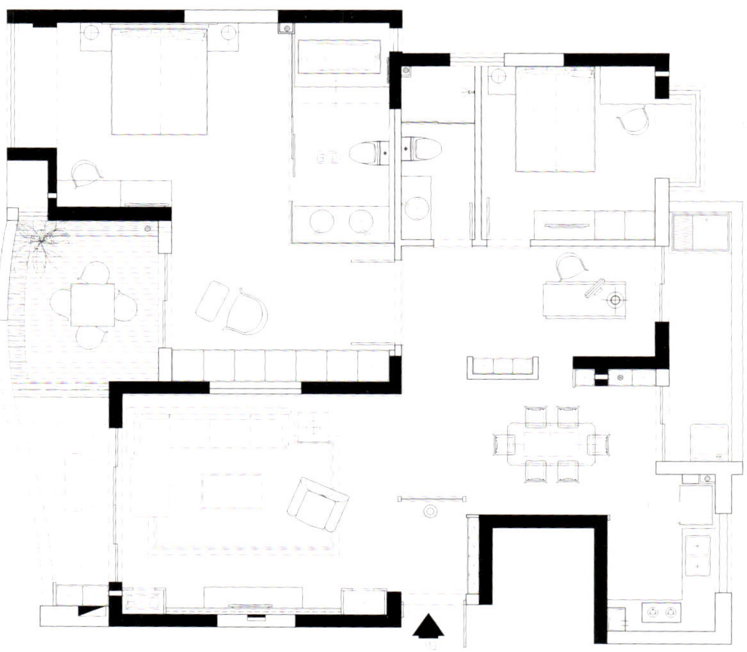

不规则的内置线性橘色光源，散发着灵动的光彩。
The irregular built-in linear orange light sources give off the bright gloss.
开敞式设计。
The open design.

三件在材质和造型上大相径庭的座椅摆放在客厅中，给人带来和谐的质感。

Three chairs which are wieldy divergent in the materials and models are placed in the living room and bring people with a harmonious quality.

电视柜上抽象的树木造型装饰品则以黑色呼应空间，以红色为之增色。

The abstract wood modeled furnishing on the TV cabinet echoes with the space by black and increases the color by red.

简单的阅读区，静谧、恬静。

The simple reading area, quiet and peaceful.

卧室设计注重柔软与温情的感受，黄色作为主色调，尽显柔和。

The design of the bedrooms emphasizes on the gentle and warm feelings. Yellow is taken as the main color with full revelation of gentleness.

中诚名仕汇 Zhongcheng Mingshihui

设计单位:福州品川设计顾问有限公司 / 面积:235 m² / 主要材料:皮革软包、清玻、铁丝网、PVC管、超白玻化砖、水曲柳染色 / 坐落地点:福州 / 工程造价:21万元 / 完工时间:2009年3月 / 摄影:吴永长

Design Unit:Pure Charm Design Consultants Limited / Area:235 m² / Main Materials:leather、clear glass、wire netting、PVC canal、glazed tiles、fraxinus mandshurica / Location:Fuzhou / Project Cost:RMB 210 000 / Finish Time:Mar,2009 / Photography:Wu Yongchang

在略显狭长的户型结构中，设计师对客厅的处理十分出彩。清雅色调勾勒空间基调，挑空的客厅十分开阔，珠帘从天而降，为空间注入婉约与柔美。星状的灯光光线在大面积竖形条纹背景墙上放纵跳舞，玻璃隔断开拓了空间的视觉美感。黑色的皮质沙发上随意搁置着白色的毛毯，与精妙的玻璃茶几构造了一个立体的几何空间。设计师利用黑、白作为连接复式空间的色调，强烈的对比满足了视觉感官的追求。在更显私密的主卧空间布置上，墙面的处理较为大胆，棕色与红色搭配的软包搭建出传统与古典风情，卫浴空间并未采取任何遮掩，玻璃的运用为生活增添了更多情趣。

In this slightly narrow and long flat-type structure, the arrangement on the living room by the designer is very outstanding. The elegant color has outlined the space mood while the overhanging living room is very wide. The bead curtain is dropping from the ceiling unexpectedly and has brought the gracefulness and softness to this space. The asteroid lamplight is dancing indulgently on the background with large area of vertical strips. The glass partition has expanded the visual aesthetics of the space. The black leather sofa is placed with the white carpet casually, which has constituted a solid geometric space together with the delicate glass tea table. The designer has applied black and white as the colors to connect the double-storey space. The strong contrast has met the pursuit of the visual organ. In regard to the layout of the more intimate master' s bedroom, the arrangement f the wall is quite bold. The brown-and-red soft package has built up a traditional and classic charm and the washroom is not covered at all. The application of glass has provided more interests to life.

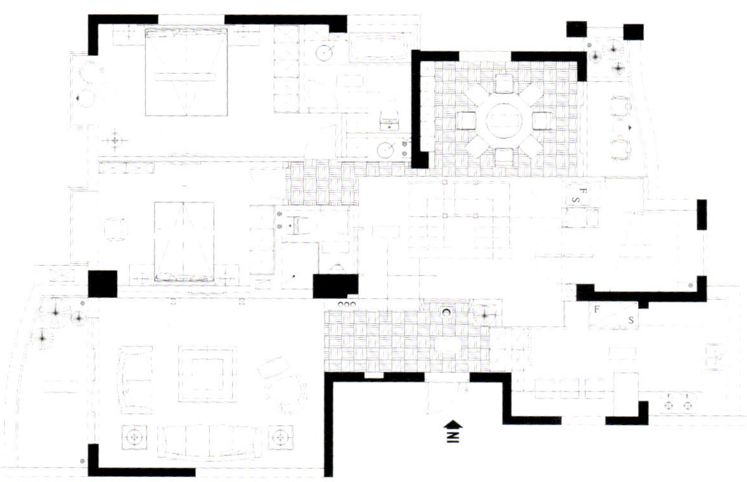

珠帘从天而降，为空间注入婉约与柔美。
The yarn curtain is unexpectedly dropping down to fill the space with gracefulness and gentleness.
挑空的客厅十分开阔。
The overhanging living room is very spacious.

书房一角。

One corner of the study room.

通透的视野。

The transparent view.

卫浴空间并未采取任何遮掩，玻璃的运用为生活增添了更多情趣。

The space of the washroom is not applied with any covering. The application of the glass has added more interests to the life.

墙面的处理较为大胆，棕色与红色搭配的软包搭建出传统与古典风情。

The wall is arranged quite bold. The soft packs in the proper combination of brown and red have built up a traditional and classic charm.

小复式–西城美墅 West Villa (Small Double Deck Space)

设计单位:于强室内设计师事务所 / 设计:于强、林志刚 / 面积:112 m² / 主要材料:石材、白色人造石、白色烤漆板、拼花马赛克、茶玻、白色成品楼梯、简欧画框、清玻、超白玻、深啡色木地板、墙纸、防腐木 / 坐落地点:杭州 / 工程造价:40万元 / 完工时间:2008年11月

Design Unit:YuQiang & Partners Interior Design / Design:Yu qiang、Lin Zhigang / Area:112 m² / Main Materials:stone、white artificial stone、white paint coating board、blossom mosaic、white end product stair、wall paper、wooden floor、syp / Location:Hangzhou / Project Cost:RMB 400 000 / Finish Time: Nov,2008

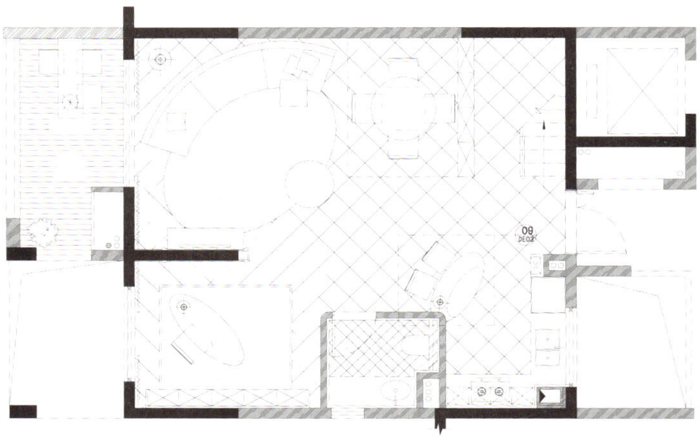

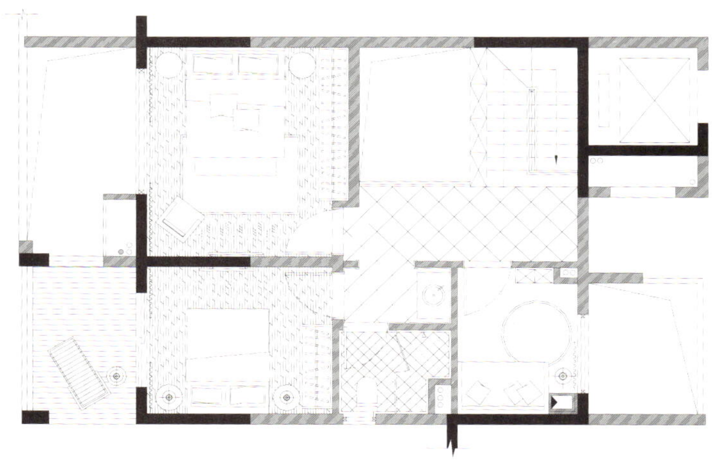

曲线造型唯美流畅。
Aesthetic and fluent curve shape.
在满足实用的基础上追求自由与开放。
In pursuit of freedom and open based on the practicability.

时尚是现代的、当下的、有时效性的、活力的象征。经典是沉淀的、稳重的、品质的、永恒的、品位的象征。活力与品味是时下高级白领阶层的生活状态与精神追求。本案的设计为他们量身定做。以现代人的眼光审视经典，以现代的设计手法与科技演义经典空间，在满足实用的基础上追求自由与开放。功能性灯光、环境光与情趣灯光多层次布置，满足不同的功能与氛围需求。饰品重于追求艺术品位、设计感与收藏感。

Fashion is a symbol of modernity, trend, timeliness, and energy. As a contrast, classics is a symbol of accumulation, steadiness, quality, eternity, and taste. Nowadays energy and taste represent the life status and spiritual pursuit of white–collar workers. This scheme is specifically designed for this group. The scheme reviews the classics from the viewpoint of modern people, and creates classic space with modern design and technology, in pursuit of freedom and openness on the basis of practicality. Multiple–level arrangements of lights with different functions are suitable for requirements of various functions and atmosphere. Decorations are focused on the quality of arts, design quality, and collection values.

原宿FLATFLAT体验店 HARAJUKU FLATFLAT

设计单位:SAKO建筑设计工社 / 设计:迫庆一郎 / 参与设计:今福佑一郎 / 面积:330 m² / 主要材料:水泥砂浆、轻钢骨构架AEP涂装、霓虹灯管 / 坐落地点:东京 / 完工时间:2008年8月 / 摄影:Chikao TODOROKI、Sasaki Studio Inc.

Design Unit:SAKO Architects / Design:Po Qinyilang / Participatory Design:Jin Fuyouyilang / Area:330 m² / Main Materials:cement mortar、lightweight steel construction、AEP coating、Neon Tube / Location: Tokyo / Finish Time: Aug,2008 / Photography:Chikao TODOROKI、Sasaki Studio Inc.

"原宿FLATFLAT"是NHN Japan公司的一家体验店。顾客可以在店内亲身体验这家公司所运营的在线游戏社区网站"Hangame"及手机游戏网站"hange.jp"提供的服务。店面位于人流不绝，尤以十几岁青少年为主的原宿明治路上。整个体验店的空间是一个幅宽3.5m、进深45m的狭长空间，而且因为反复的增建、修建，形成了目前复杂的平面结构。

"原宿FLATFLAT"的设计中提出了"未来公园"的理念，即在任何人都可以轻松进入且自由活动的现实空间里配合上虚拟的要素，是一个"有机的wet"与"被漂白过的dry"共存的空间，表示着在现实与虚拟两义性上运营的现代社会。

由曲线构成的"有机的wet"具有洞窟般的空间特性，随着场景的不断展现，顾客可根据个人的目的或心情找到最适合自己的地方，展现出一个多样的空间。基于人体尺寸而设定的曲线形成有机的形态，引导人们自然而然地与之接近和亲近。同时由白墙、天花板的霓虹灯管、水泥砂浆地面的无机的质感构成的"被漂白过的dry"产生出虚拟的特性。为了防止水泥沙浆地面出现裂缝而故意做出的缝隙，以及霓虹灯的线条都与墙面的曲线一致。由此产生的相辅相成的多层次效果，刺激着来访者的探求心，将其引向细长空间的深处。

"HARAJUKU FLATFLAT" is an experience shop of NHN Japan Company. It is a 3.5m wide and 45m deep narrow and long space and has formed the current complicated plane structure due to the repetitive expansions and modification.

The design has proposed the philosophy of "Future Garden", which means in the real space where anyone can easily enter and take activities freely with arrangement of the virtual elements, an "organic wet" and "bleached dry" coexistent space and delivers a modern society running in both reality and virtuality.

The "organic wet" space constituted by curves has the cavern-like space features. With the continuous display of the scenes, the customers can find a most favorable place for themselves according to the individual purposes or moods. The gaps that are made deliberately in order to prevent the cracks on the concrete floor and the lines of neon lights are consistent to the wall curves. Their subsequent multi-level effects set off each other, inspire the exploring curiosity of the visitors and lead them to the depth of this narrow long space.

曲线构成的"有机的wet"具有洞窟般的空间特性。
The "organic wet" cave-like space characteristics constituted by the curves.
白墙、霓虹灯管、水泥砂浆产生出虚拟的特性。
The dummy characteristic generated by the white wall, neon light tubes and cement mortar.

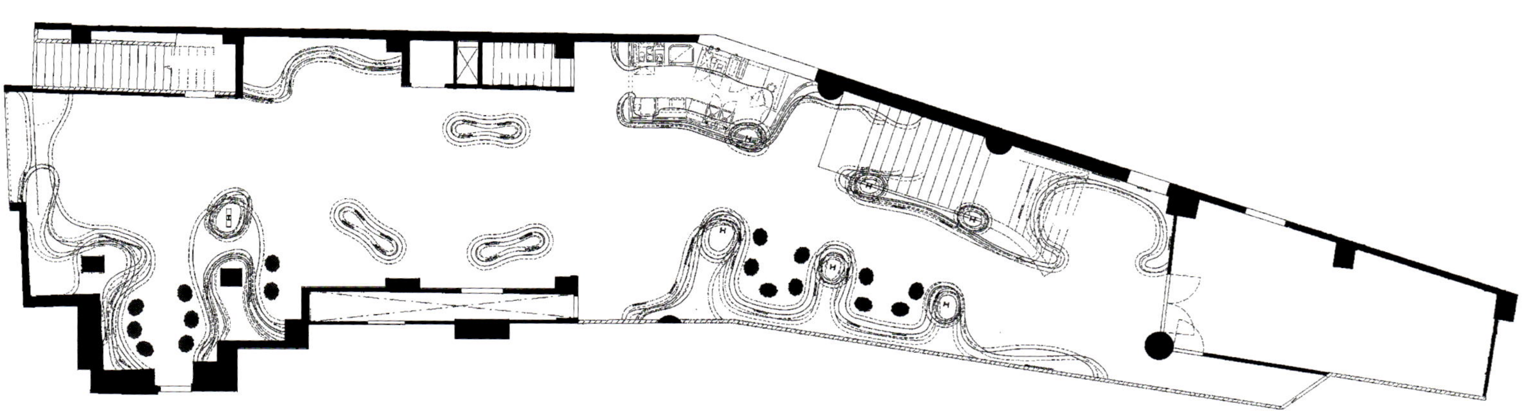

Cafe
Touch Panel
PLAY IT NOW

随着场景的不断展现，顾客可根据个人的目的或心情找到最适合自己的地方。

The customers can find the most suitable place for themselves according to the individual purpose or mood with the continuous display of the scenes.

基于人体尺寸而设定的曲线，引导人们自然而然地与之接近和亲近。

The ergonomic curves lead people naturally to get close to and intimate with them.

为防止水泥沙浆地面出现裂缝而故意做出缝隙。

The gaps that are intentionally made to avoid the cracks on the cement mortar ground.

Miss Zhang精品服装店 Miss Zhang's Boutique

设计单位:杭州正午装饰设计咨询有限公司 / 设计:顾吉 / 面积:200m² / 主要材料:绒布、木材、墙纸 / 坐落地点:杭州 / 工程造价:50万元 / 完工时间:2008年9月 / 摄影:陈杰

Design Unit:C+ZZ international design company limited / Design:Gu ji / Area:200m² / Main Materials: flannel 、wood、wallpaper / Location: Hangzhou / Project Cost:RMB 500 000 / Finish Time:Sep,2008 / Photography:Chen jie

Miss Zhang坐落于西湖边一条幽静的路边，两层楼的小店更像是一座服饰艺术品收藏家的房子，以优雅的环境来提供服饰艺术品与宾客们的亲密交流。站在店外，就已经被店里奢华的味道深深吸引，金影斑驳的橱窗内陈列出的是精致华丽的衣物，奢华的古典欧式风格四处可见。复古图案的欧式墙面、华丽的衣物橱窗、花样年华似的温暖光线、贴心的休憩空间，细节上的装饰无不表达出奢华的内蕴。那种欧洲特有的完美气质和艺术精华流淌在每样精雕细琢的家具之上。转角楼梯上去墙面全部采用紫灰色的砖石绒墙面，在灯光映射下泛出淡紫色光泽，给人华贵、幽婉的气质。

Miss Zhang is located at a sequestered roadside beside the West Lake. This two-story store is more like a house of a dressing artwork collector and provides the intimate communication between the dressing artworks and the visitors with an elegant environment. Standing outside the store, you have been deeply attracted by the luxurious flavor in the store. What are displayed in the shop window with motley golden shadows are the delicate and magnificent dresses with full revelation of a classic continental style everywhere: the archaic patterned continental style wall, the luxurious shop window, the flowery warm light and the intimate rest space, of which the detailed decorations have delivered a luxurious meaning without exception. That unique perfect European bearing and the artistic essences are flowing on each piece of exquisitely carved furniture. The wall above the turning staircase is completely covered with the violet gray brick amianthus, which emits the light violet gloss under the reflection of the light and delivers a luxurious and deep atmosphere.

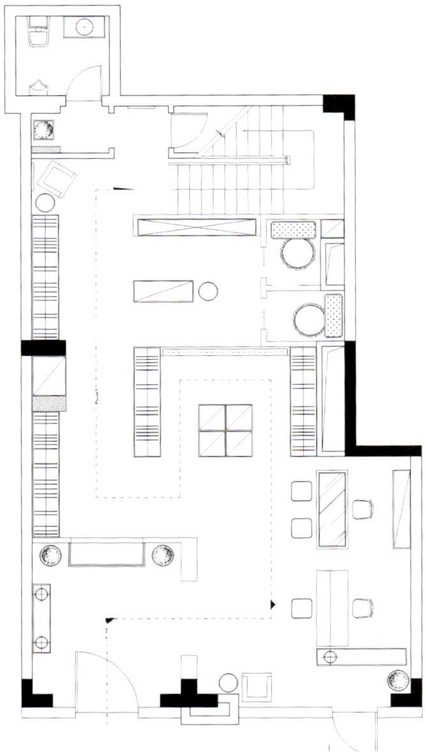

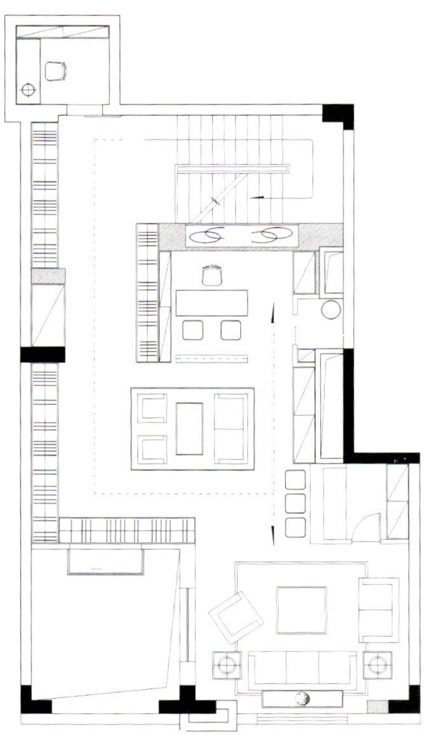

小店更像是一座服饰艺术品收藏店。
The shop looks more like a collection shop for clothing artworks.
华丽的衣物橱窗。
Luxurious Clothing Shop Window.

GIORGIO ARMANI

细节上的装饰无不表达出奢华的内蕴。
The detailed decorations are invariably delivering a luxurious meaning.
站在店外，就已经被店里奢华的味道深深吸引。
Standing outside the shop, you have been deeply attracted by its luxurious atmosphere.
复古图案的欧式墙面。
The antique patterned continental-style wall.
花样年华似的温暖光线。
The blooming warm beams.

空间里的盒子 A Box in a Space

设计:林开新 / 面积:130m² / 主要材料:水泥地板、水泥漆、防火板、写真图片、背漆玻璃 / 坐落地点:福州 / 工程造价:8万元 / 完工时间:2008年7月 / 摄影:吴永长

Design:Lin Kaixin / Area:130m² / Main Materials:cement floor、cement paint、board fireproofing、picture、back-painted glass / Location:Fuzhou / Project Cost:RMB 80 000 / Finish Time:Jul,2008 / Photography:Wu Yongchang

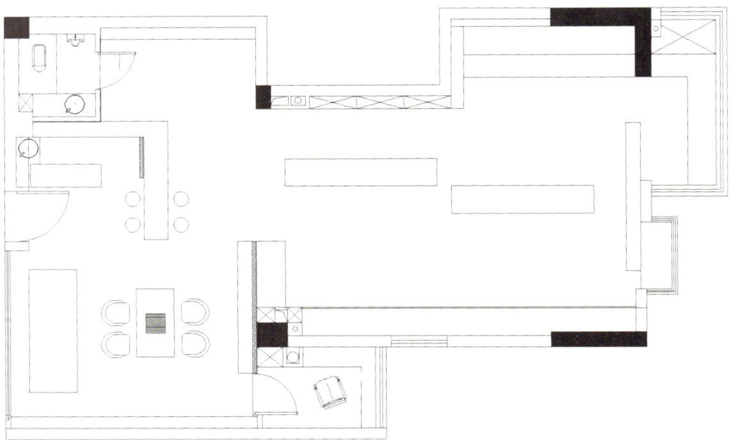

这是一个经营设计家具及设计小品的展示厅。以盒子为设计元素，把盒子去边、加边、镂空留边等手法，设计成不同产品的展示柜及工作台。这些盒子在特定的位置相互冲插，并与大空间的结构相呼应，形成一种空间元素的统一协调。大空间盒子用平面图片的形式来表示：图片内容是建筑师、设计、设计产品等，和天花垂吊的图片隔断形成统一呼应，从而传达出设计与时尚的关系。

It is an exhibition hall engaged in designed furniture and small articles. A box is taken as a design element. By removing, adding, hollowing out and reserving the edges of the boxes etc., the boxes are designed into the exhibition cabinets and working tables for different products. These boxes are interwoven together in the specific positions and echo with the structure of the big space so that a unification and harmony of the space elements is formed. The boxes of big spaces are expressed through the forms of plane pictures: the contents of the pictures include the architects, designs, designed products etc., which form into a unified echo together with the picture partition down from the ceiling to deliver the relationship between the design and fashion.

别具一格，顶、墙面都成了展厅的主角。
In a unique style, the ceiling and wall play a leading role in the exhibition hall.
家具设计师的写真图片也不失为很好的装饰元素。
The portraying pictures of the furniture designer are also a good decorative element.

Tags
and the City
涂鸦与城市
DISERTACI

Coniva橱柜展厅 Coniva Cabinets

设计单位:珠海市空间印象装饰设计工程有限公司 / 设计:霍承显 / 参与设计:彭昀 / 面积:360 m^2 / 主要材料:抛光砖加工、乳胶漆、肌理涂料、镀锌板烤漆、黑色烤漆玻璃、工艺吊灯 / 坐落地点:珠海 / 工程造价:32万元 / 完工时间:2009年6月

Design Unit:SIDD / Design:Huo Chengxian / Participatory Design:Peng yun / Area:360 m^2 / Main Materials: polished tiles、emulsion paint、paint、galvanized plate、storing varnish glass series / Location:Zhuhai / Project Cost:RMB 320 000 / Finish Time:Jun,2009

绿色主题的sofa、AD设计在深灰色的墙体、白色的地板映衬下，营造一种时尚风。拦而不封的装饰墙令空间尽显神秘感。相对独立的洽谈选材区域，减少过客干扰，同时提供便利的临时设计平台。旁边，设置了体验区，同时兼具提供饮品和小食的服务功能。天花独辟蹊径地将白色悬板与黑色铁艺灯搭配，组合时尚。

灯光设计别致有加，装饰灯造型灵感来源于中国传统的灯笼，悬挂于天花上的射灯围成方形，烘托整个展厅的氛围。简单的材料和造型应用，将具有浓厚传统的细致做工与现代时尚元素完美融合，彰显时尚感与经典文化。

The sofa subjected as green and AD design have created a fashionable style under the setoff of the dark gray wall and white floor board. The decorative wall that is blocked but not sealed makes the space extremely mysterious. The comparatively independent discussion and material area has reduced the interference from the passers-by and also provided a convenient temporary design platform. An experience area is arranged beside it, also providing drinks and snacks. In the ceiling, the white suspension boards are uniquely and specially arranged with black iron a lamp, which looks very fashionable.

The light is designed extremely unique. The model of the decorative lamp is inspired by the Chinese traditional lantern while the suspended spot lamp on the ceiling is surrounded into a square to set off the atmosphere of the whole exhibition hall. The application of the simple materials and models has perfectly integrated the strong traditional exquisite handcraft work together with the modern fashionable elements, prominently showing the fashionableness and classic cultures.

白色地板营造一种时尚风。
The white floor board creates a fashion style.
不经意的细节体现细致的表现。
The unintentional details are reflecting the delicate demonstrations.

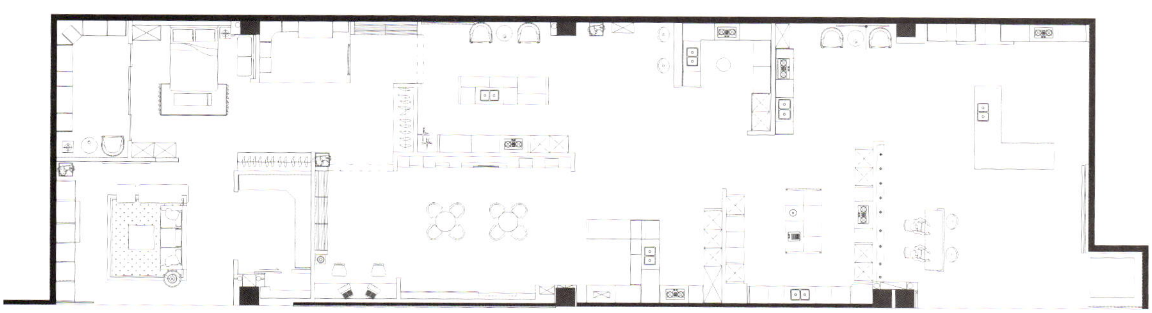

any Profile
004, Zhuhai Coniva Furniture Co.
esign, development, produ and
l cabinets, wardrobes and oratory
d office and production are both
Industrial Zone with scenery.
overs a land area of 20 00 square
e area of its factory being
eters, and has an annu production
f more than 5, sets.
s managers built up brand
g Coniva' s entry into y of
-end cabinets and wa
al and manpower hav en in ted
anagement, produc ation and
ation to provide the omers with
healthy high-grade products
in quality and affor in prices.
of development strategic
ive, a leading co

天花独辟蹊径地将白色悬板与黑色铁艺灯搭配，组合时尚。
The ceiling has uniquely arranged the white suspended boards together with the black ironwork lamps in a fashion way.
拦而不封的装饰墙令空间尽显神秘感。
The decorative wall that is obstructed but not sealed makes the space mysterious.
绿色主题的sofa位于洽谈选材区。
The sofa with a green theme is located in the discussion and material selection area.

coniva康尼华

炜俊浴室设备有限公司 Weijun Shower Equipment Co., Ltd.

设计单位:墨象·奇思度设计策划有限公司 / 设计:梁宇曦 / 面积:800 m^2 / 主要材料:铝塑板、工艺墙纸、透光膜、乳胶漆、不锈钢、马赛克、钢化玻璃 / 坐落地点:佛山 / 工程造价:150万元 / 完工时间:2009年2月 / 摄影:梁宇曦

Design Unit:MOREX design consulting / Design:Liang Yuxi / Area:800 m^2 / Main Materials: ACP、crafts wallpaper、UHTC、emulsion paint、stainless steel、mosaic、tempered glass / Location:Foshan / Project Cost:RMB 1 500 000 / Finish Time:Feb,2009 / Photography:Liang Yuxi

本案旨在结合展厅的定位创造富有古典与后现代主义结合的空间，采用叠加、共处等手法，大胆运用空间概念，结合产品切入主题。

总部一、二层用于展示产品，三层用于办公。在首层设计中，采用不规则线条，高纯度对比色打破常规，把古典主义组合在新的情景中。空间有层次感，处处讲究古典主义的华丽和后现代主义的简洁大方。二层设计延续首层设计理念，在空间上创造一种感性与理性、传统与现代的环境。

This design is tended to create a space of classism and postmodernism in combination with the orientation of the exhibition hall. The techniques of stacking and coexisting etc. are applied and the space conception is boldly used in combination with the products to match the subject.

The first and second floors of the headquarters are used to exhibit the products while the third floor is used for office. In the design of the first floor, the irregular lines are applied while the high pure contrasting colors have broken through the convention and the classism is assembled into the new scenery. The space has the sense of levels and attentions are paid to the magnificence of the classism as well as the simplicity and grandness of the postmodernism everywhere. The design of the second floor has continued the design philosophy of the first floor to create a sensible and rational, traditional and modern environment from this space.

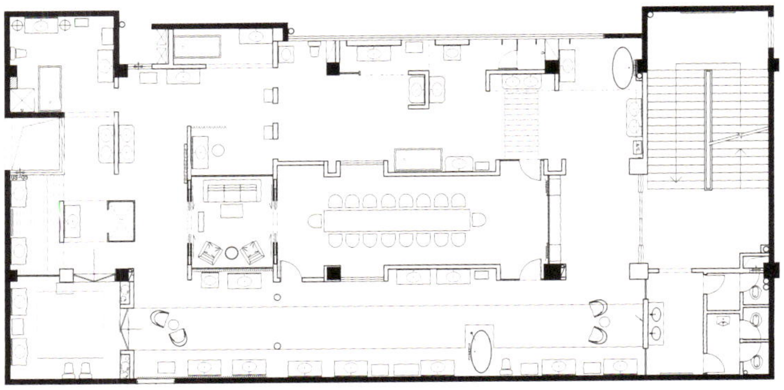

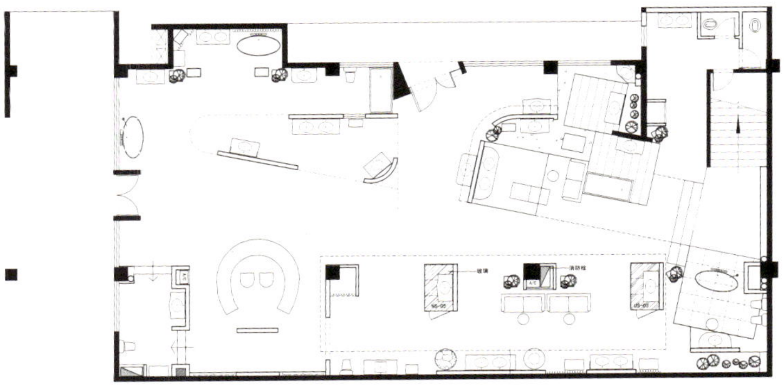

古典与后现代完美结合。

The perfect combination between the classism and post-modernism.

各功能空间合理贯通。

Each functional space is properly connected with each other.

静谧温馨的一隅。
One quiet and warm corner.
顶面与地面黑白极致对比。
The extreme contrast between the black ceiling and white ground.
高纯度对比色打破常规。
The high pure contrasting colors have broken the conventions.
采用叠加、共处等手法，大胆运用空间概念。
The techniques such as piling and coexisting are applied and the space conception.
古典主义被组合在新的情景中。
The classism is combined into the new scene.

赛德斯邦陶瓷展厅 Cerlords Ceramics

设计单位:东莞市形上几何室内设计有限公司 / 设计:卢迅 / 参与设计:林涛、王永元、李文尧、罗洁 / 面积:4000 m² / 主要材料:仿古砖、透光膜、钢化玻璃、拉丝不锈钢、银镜、槽钢、磨砂有机片、乳胶漆 / 坐落地点:佛山 / 工程造价:280万元 / 完工时间:2008年10月 / 摄影:高佬

Design Unit:XingShangJiHe Design / Design:Lu xun / Participatory Design:Lin tao、Wang Yongyuan、Li Wenyao、Luo jie / Area:4000 m² / Main Materials:rustic tile、UHTC、tempered glass、stainless steel、silver mirror、emulsion paint / Location:Foshan / Project Cost:RMB 2 800 000 / Finish Time:Oct,2008 / Photography:Gao lao

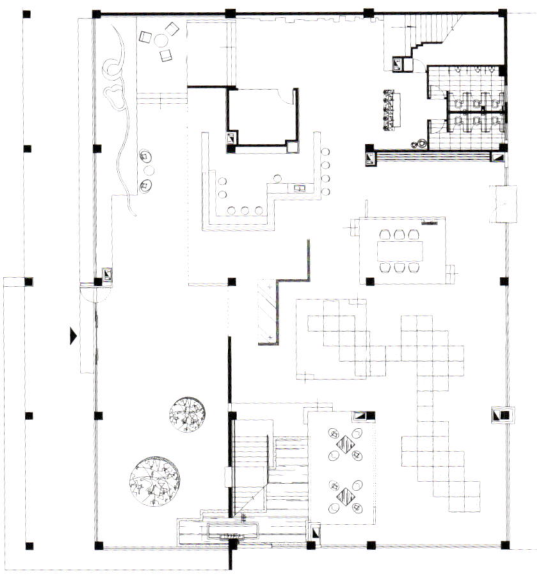

这是一个充分体现“自然、生态、环保”的现代陶瓷品牌体验中心。空间充分体现品牌的核心思想，用后现代主义色彩感知时尚脉搏，以极简的空间结构创造核心竞争力的唯一性。在意化实物的同时将品牌、思想、意念物化下来，谱写出赛德斯邦自己的砖道。设计师在展示中解决的不是砖的问题，而是消费者的态度、观点、情感和审美方式等。

外墙大量应用墙漆、玻璃等材料，追求情景交融的生态景观，使城市与大自然融合在一起。构造一条长廊来诠释品牌内涵，突破限制，与其“让世界重新想象”的创意相吻合。哪怕是一个角落，也要让消费者体验到对生活的态度和人生的价值，给人以想象。

It is a experience center of modern ceramic brand with full expression of “nature, ecology and environmental protection”. The space has sufficiently reflected the core philosophy of this brand – by using the postmodernism colors to feel the fashion trends and creating the uniqueness of the core competitive power through the extremely simple space structure. The brand, philosophy and ideas are materialized while idealizing the material objects to compose a tile path of Cerlords’. The designer shall settle the problems of the customers’ attitudes, opinions, emotions and aesthetic method etc. instead of the tile issues in the exhibition.

The exterior wall has applied a great quantity of materials such as wall paints and glass etc. in pursuit of an ecologic landscaping with the integration of feelings and sceneries so that the city and the nature can be integrated together. A long corridor is built up to interpret the brand meanings, break through the restrictions and make it match with its innovation – “Make the world re-imagine”. Even a corner shall make the customers experience the attitude towards life and life value and leave people with imagination.

特立独行的球型展示方式。
The noteworthy and independent ball-shaped display.
构造一条长廊来诠释品牌内涵。
One long corridor is built up to interpret the meanings of the brand.

交错折叠，令空间变幻无穷。
The staggering, folding and overlapping make the space with infinite changes.
自然的枝桠团簇在一起。
The natural branches and forks are clustering together.
偌大的空间安排得周到、合理。
Such a big space is arranged considerate and reasonable.
层次错落的立柱既分隔了空间，又展示了产品。
The orderly strewn vertical columns not only separate the space, but also display the products.

陶瓷城JNJ专卖店 JNJ Franchised Store in Ceramic City

设计单位:佛山市城匠建筑设计院有限公司明威分公司 / 设计:谢智明 / 参与设计:霍律鸣 / 面积:155m^2 / 主要材料:JNJ玻璃马赛克、玻璃、不锈钢 / 坐落地点:佛山 / 工程造价:50万元 / 完工时间:2009年5月 / 摄影:钱翔

Design Unit:MingWay Building Design Engineering Limited / Design:Xie Zhiming / Participatory Design: Huo Lvming / Area:155m^2 / Main Materials:JNJ glass mosaic、glass、stainless steel / Location:Foshan / Project Cost:RMB 500 000 / Finish Time:May,2009 / Photography:Qian xiang

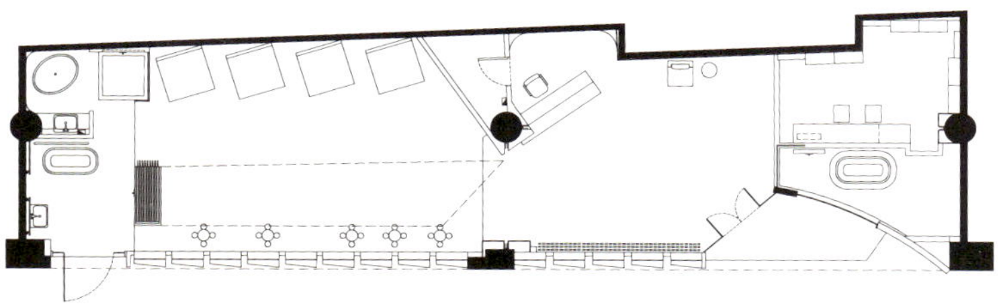

JNJ Mosaic专卖店保留JNJ专卖店一贯概念化设计思想，同时为推广个性零售的功能，在空间内多方位裁剪了个性板间，合理利用视角空间关系，把多个夺目的马赛克图案有机融合，整体感觉还显简洁，为本案设计的关键。成功的专卖店应在保留品牌风格的同时兼容美观与实用，达到完美的平衡。

JNJ Mosaic franchised store has kept the consistent conceptualized design philosophy of JNJ franchised stores. Furthermore, to promote its individualized retail function, the individualized sample rooms are divided in various ways in the space. The relationships among the visual spaces are reasonably applied to organically integrate the various brilliant mosaic patterns together so that the entire looks quite simple and concise, which is the key to this design. A successful franchised store shall keep the style of the brand while comprising the beautifulness and practicality to reach a perfect balance.

空间内多方位裁剪了个性板间。
The individualized sample rooms are divided in various ways in the space.
灯光不自觉地将人的视线引到产品本身。
People' s sight lines are naturally attracted to the products.

BOMY
PRESTIGE
港澳
名品
gangaomingpin

BOMY

女装与男装分列上下两层。

Women's wear and men's wear are separately arranged in the upper and lower layers.

镜面的圆顶设计扩展了空间。

Round top design of the mirror expands the space.

楼梯旁的小景活灵活现。

Lifelike small scene beside the stairs.

收银台简单、大方。

Simple and decent cashier's desk.

卡德维展厅 Kaldewei Exhibition Hall

设计单位:光线装饰设计工作室 / 设计:许志国 / 面积:120 m² / 主要材料:烤漆玻璃、有色乳胶漆、细钢管、实木地板、地砖、不锈钢、白色软膜 / 坐落地点:南京 / 工程造价:15万元 / 完工时间:2008年11月 / 摄影:文宗博

Design Unit:Light Decorative Design Studio / Design:Xu Zhiguo / Area:120 m² / Main Materials:painted glass、emulsion paint、steel tube、solid wood floor、floor tile、stainless steel、white soft mask / Location:Nanjing / Project Cost:RMB 150 000 / Finish Time:Nov,2008 / Photography:Wen Zongbo

本案摈弃了一般卫浴产品展示中单元间情景再现的手法，把普通系列打散后重新排列组织，或聚或散。新的空间流线顺势而生，结合店面运营需要，安排重点产品，以达到类似橱窗展示吸引眼球的效果。材料以深灰色涂料结合原木地板为主，局部搭配烤漆玻璃及白色细钢管，在突出产品的同时，增加展厅的现代气息及趣味性。

The scheme discards inter-unit scene reappearance during the regular demonstration of bathroom products, and rearranges regular series after breaking them up, thus creating a flexible combination of gathering or scattering. New space lines emerge accordingly to convenient arrangement of key products in accordance with operation requirements. The purpose is to attract customers, just like similar shop windows do. The materials mainly include deep grey paint plus hardwood floors, with lacquered glass and white thin steel tubes partially equipped. This highlights the products and makes the exhibition hall more modern and interesting.

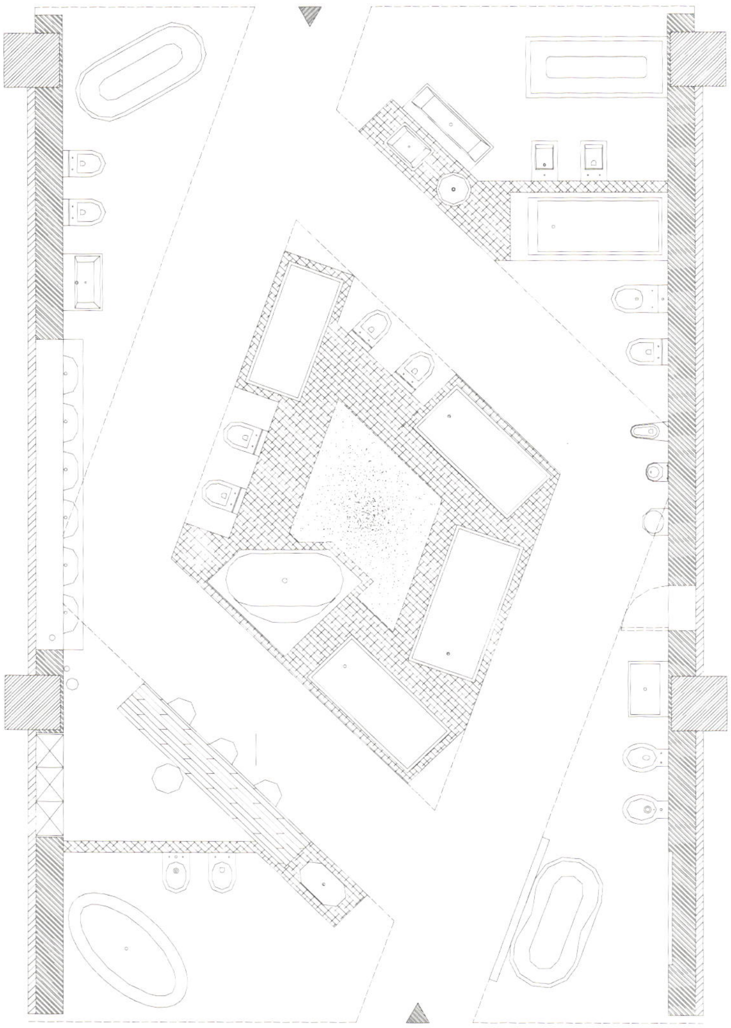

深灰色涂料上打上展厅名字，沉稳、内敛。
Name of the exhibition hall is on the dark grey painting, represented in a steady and convergent background.
不同材质区分了过道和展区。
The corridor and the exhibition zone are separated by different materials.

CATALANO
卡特雷诺 意大利

隔断中间断开，放上配饰打上灯光。
Partition walls are separated in the middle by decorative lamp light.
精致的细部设计。
Elaborative detail design.
普通系列打散后重新排列组织，或聚或散。
Rearranged and reorganized common series, gathered or scattered.
重点产品重点展示。
Exhibition for the key products.

KALDEWEI
GEBERIT

木亿家意大利家具展厅 Muyijia Italian Furniture Showroom

设计单位:大木设计中国发展有限公司（北京公司）/ 设计:唐凡 / 参与设计:赵家林、刘德伟 / 面积:1200 m² / 主要材料:乳胶漆、丰镇黑花岗岩 / 坐落地点:北京 / 工程造价:120万元 / 完工时间:2009年4月 / 摄影:唐凡

Design Unit:Damu Design China Developing Corporation(Beijing) / Design:Tang fan / Participatory Design: Zhao Jialin、Liu Dewei / Area:1200 m² / Main Materials:emulsion paint、fengzhen black granite / Location: Beijing / Project Cost:RMB 1 200 000 / Finish Time:Apr,2009 / Photography:Tang fan

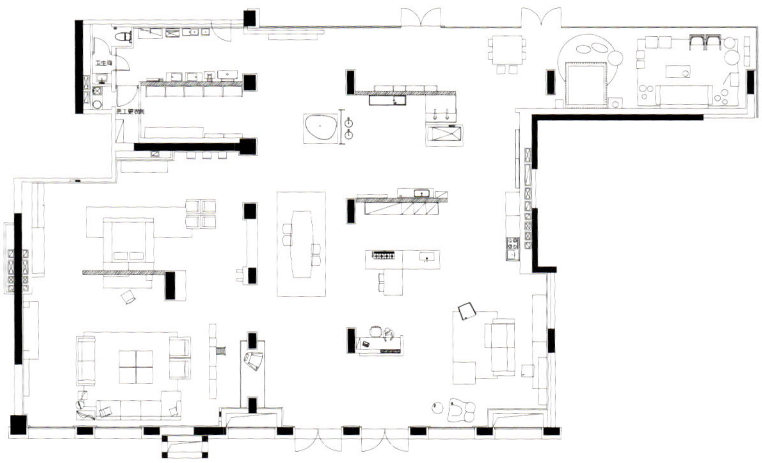

在本案的设计中，主要精力都在空间的布局与光环境的设计中。合理地划分空间，强调家具与空间的关系，注重灯光在产品与空间中的作用是本案的设计宗旨。减掉一切多余的装饰，使空间变得纯净。采用最简洁的形式和最普通的装饰材料来营造一个家具秀的舞台，突出家具产品本身的张力与品质，使其成为空间中真正的主角。

Focus on the space layout and the optical design for environment. The aims of this design are to divide the space reasonably, to emphasize on the relationship between furniture and space, and also on the light's role in the products and the spaces. The spaces become pure by taking off all extra decorations. The most concise form and the most common materials together create a furniture show stage which can highlight the original tension and quality of the furniture so that it become a leading role in the spaces.

合理地划分空间，强调家具与空间的关系。
The design divides the spaces reasonably and focuses on the relationship between the furniture and the spaces.
减掉一切多余的装饰，使空间变得纯净。
Taking off all extra decorations , the spaces become pure.

注重灯光在产品与空间中的作用。
highlight the role of the light between the products and the spaces.
通过半封闭式隔墙或者玻璃划分出一个个小展示空间。
The semi-closed walls or glasses divide out numbers of small exhibition space.
家具本身成为空间中真正的主角。
The furniture become a true leading role in the spaces.
家具的深色与墙地面的浅色对比。
The dark paint on the furniture provides a contrast for the light color of the wall and floor.
采用最简洁的形式和最普通的装饰材料营造家具秀的舞台。
The most concise shape and the most common decorative materials together create a furniture show stage.

谷腾设计师工作室 Guteng Designer's Studio

设计单位:谷腾设计师工作室 / 设计:谷腾 / 面积:140m² / 主要材料:玻璃、镜面、不锈钢 / 坐落地点:上海 / 工程造价:26万元 / 完工时间:2009年5月 / 摄影:谷腾

Design Unit:Guteng Designer' s Studio / Design:Gu teng / Area:140m² / Main Materials:glass、mirror surface、stainless steel / Location:Shanghai / Project Cost:RMB 260 000 / Finish Time:May,2009 / Photography:Gu teng

此空间的格局进行了较大的改造。由于面积较小，所以靠设计实现了很多功能的要求，在满足使用功能的基础上加以视觉感受，运用了镜面和玻璃的特性使其空间从视觉上加大了延伸感和趣味性，二层的搭建与玻璃的斜线的结合，使空间的体量加大，而且对于没有围栏的过道也起到安全保护的作用。家具的设计采用了一种元素多变的形式，视觉上更统一简约。

The layout of this space has been modified greatly. As the area to be used is quite small, its design has made up many practical functions, which not only meet the functional requirements, but also is added with the visual feelings. The properties of the mirror and glass are applied to visually expand its space and increase its interest. The buildup of the second floor and the combination of the inclining lines of the glass has enlarged the block of the space. Furthermore, the design of the furniture that functions as a protection to the corridor without railing has applied a form of variable elements so that this space looks more uniform and concise.

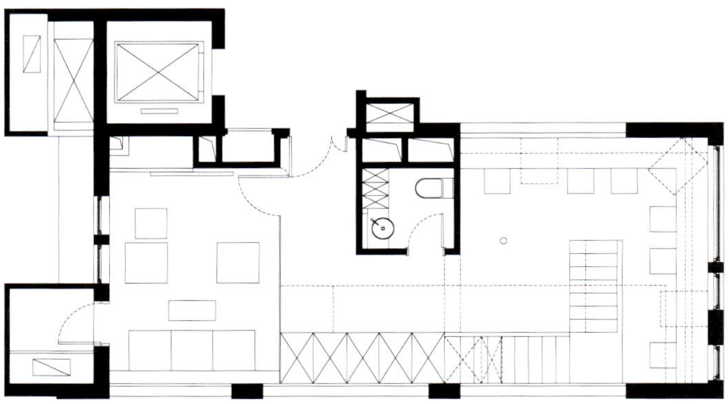

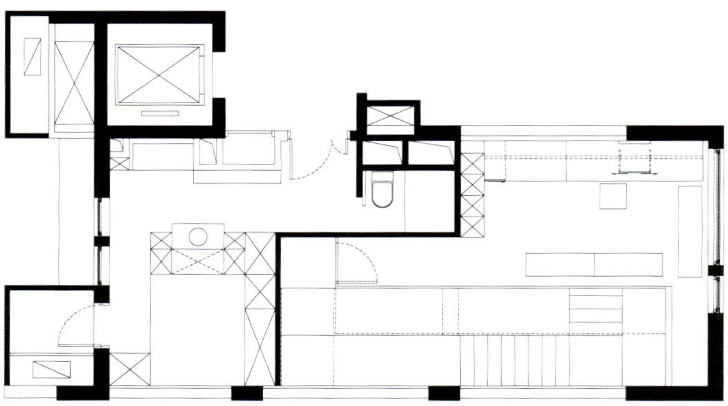

独特的墙面造型。
Unique Wall Model.
交错的光影效果富有趣味性。
Staggered light and shadow effects are interesting.
透明纯净的空间。
Transparent and Pure Space.

从二层俯视一楼。
Looking down the ground floor from the second floor.
办公家具的设计采用同一种元素而多变的形式。
The design of office furniture applies the changeable forms in one element.
不规则白色造型体块别具匠心。
Irregular white model blocks have a unique style.

大面积镜面的反射增加了延伸感。
The reflection of large area of mirror increases the extensional sense.
二层的搭建和玻璃斜线的结合，加大了空间体量。
The buildup and combination of glass diagonals at the second floor have enlarged the space.
白黑银的色彩搭配相得益彰。
The color arrangement of white, black and silver match with each other perfectly.

美国WAC灯具工厂总部 WAC Fixture Factory Head Office

设计单位:崔华峰空间设计艺术顾问工作室 / 设计:崔华峰、江健祥、林江峰、胡巍挺、范慧 / 面积:15000 m² / 主要材料:瓷砖、千思板、乳胶漆、实木 / 坐落地点:东莞 / 工程造价:480万元 / 完工时间:2008年3月 / 摄影:江健明

Design Unit:Cui Huafeng Space Design Art Consultant Studio / Design:Cui Huafeng、Jiang Jianxiang、Lin Jiangfeng、Hu Weiting、Fan hui / Area:15000 m² / Main Materials:tile、trespa、emulsion paint、solid wood / Location:Dongguan / Project Cost:RMB 4 800 000 / Finish Time:Mar,2008 / Photography:Jiang Jianming

本案为美国WAC灯具中国工厂总部，顶部是光学术交流中心。当设计完成时，我们意识到——真正的设计师是上天，上天赋予了空间充满朝气。开业那天：白色的空间中，清澈的阳光，深色着装的贵客们摇着香槟。

他们问：灯去哪了？

我们答：灯给上天没收了！

你们做了什么？

我们在上天的指引下为空间安排了人类的空间艺术品！

天啊！

This case is WAC China manufacture head office,15000 m²,the top is optical academic exchange center.

When the case is done, we sensed that the real designer is God, it is God that could make this mortal a wonder.

At opening day, honoured guests enjoy the champagne with the sunshine wraped around.

"where the lamp-house?" guests ask

We say: God take it back!

"how do you achieve this?"

"we just follow God's devree arranged a artwork for this space"

"oh God!"

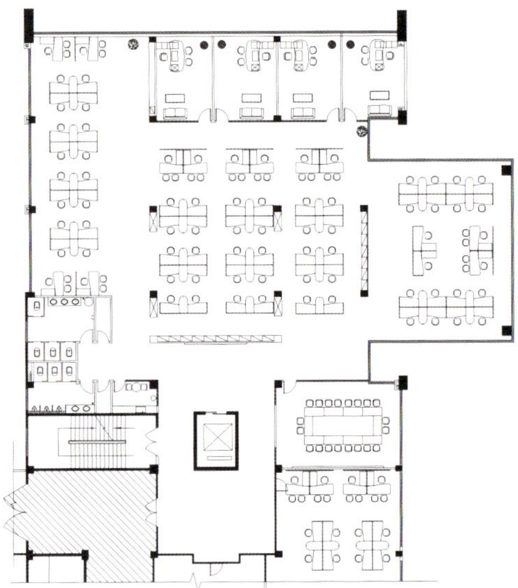

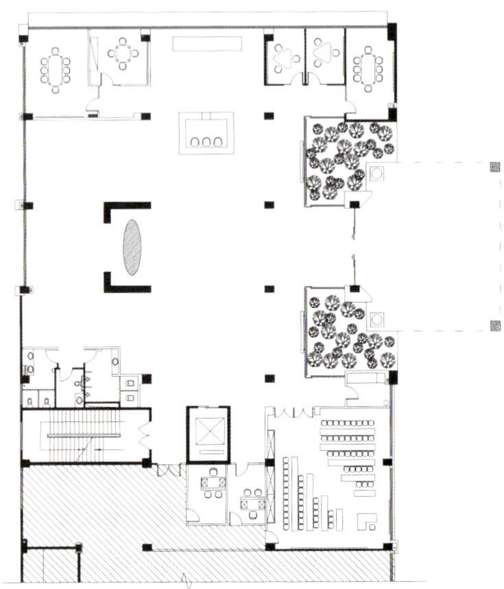

尽享自然的阳光。

Fully enjoy the natural sunlight.

光与影构成的小魔术。

A small magic constituted by light and shadow.

生动的人物配饰平添趣味。
The vivid personal ornaments naturally increase the interests.
白色立体的空间，人造光源在此点缀。
A white three-dimensional space, dotted by artificial light sources.
形似于撒满阳光的天井。
Similar in shape to a dooryard full of sunlight.
阶梯形配饰成为视觉焦点。
The step-type ornaments become the visual focus.

渠成国贸办公空间 Qucheng International Trade

设计单位:宁波市高得装饰设计有限公司 / 设计:范江 / 参与设计:郑磊 / 面积:1150 m^2 / 主要材料:白玉纹大理石、古铜色不锈钢、EV银橡、白洞石、烤漆玻璃、软木饰面、彩色乳胶漆、塑胶地板 / 坐落地点:宁波 / 工程造价:100万元 / 完工时间:2009年1月 / 摄影:文宗博、潘宇峰

Design Unit:Gotel Design Studio Limited. / Design:Fan jiang / Participatory Design:Zheng lei / Area:1150 m^2 / Main Materials:marble、stainless steel、silver oak、travertine white、storing varhish glass series、wood veneer、emulsion paint、plastic floor / Location:Ningbo / Project Cost:RMB 1 000 000 / Finish Time:Jan,2009 / Photography:Wen Zongbo 、Pan Yufeng

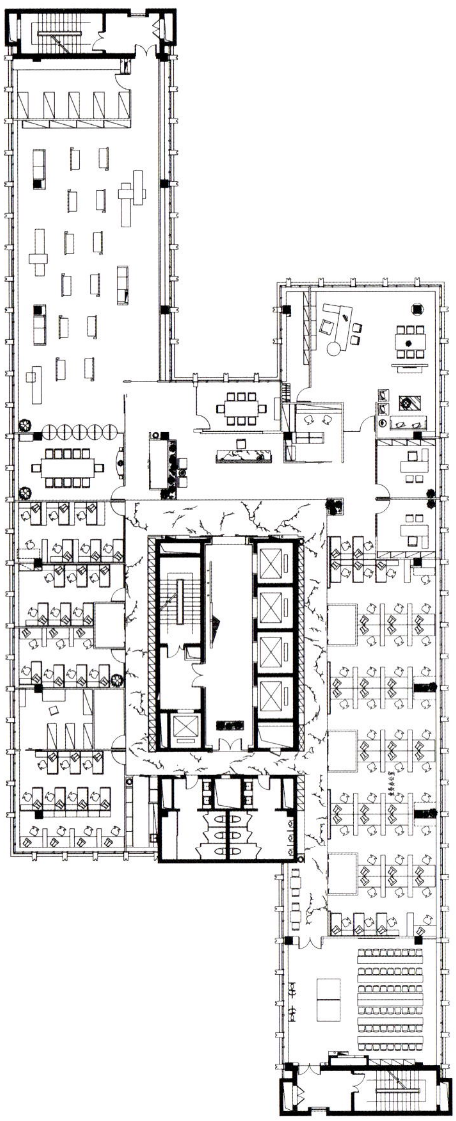

地面及墙面的白玉纹大理石凸显品质。

The white jade lined marble on the ground and wall has conspicuously shown the space qualities.

前台空间稳重大气。

The reception space looks calm and grand.

这是与业主的二度合作。办公空间位于高品质的写字楼里，占整一层。国际化、大气，但不走时尚前沿路线是此次设计的定位。建筑空间比较狭长，办公室为两种形式，一种是单间、一种是大办公区域。大办公区里设计了几个岛屿式的储藏室，使储藏更为集中并把空间围合成半开放式，办公楼采光较好但展示厅不需要很多的自然光，就用整条的展板进行遮蔽，通过灯光使展示厅显现戏剧化效果。

“渠成”取水到渠成之意，所以在前厅特意设计了一个造型，从顶部悬挂下来刻有水波纹的玻璃造型，对应着下面小水池“水到渠成”的印章，体现企业顺其自然的经营理念。凸现品质还体现在材料的使用上，如云水纹般的白玉纹大理石、镀铜不锈钢板、烤漆玻璃、塑胶地砖……配合大块面造型及木切片与玻璃的应用使空间现代含蓄但不冷峻而且有内涵。展示厅跳跃的橘红色是他们的标志色，与中间展架蓝色的灯光形成一种对比。展示厅把握的基调略有不同，令空间有变化、有起伏。

设计贴近自然，始终以人为本。

Internationalization and grandness, but without pioneer fashion are the orientation of this design.

The building space is narrow and long and the offices have two types: single room and opening office area. Several island-type storage rooms are designed in the opening office area so that the storage can be more centralized and the space is enclosed into a semi-opening style. The exhibition hall has applied the whole exhibition board to block the natural light so that the lamplight looks more dramatic.

The front hall has been specially designed with a model. The glass model carved with water waves hung from the ceiling echoes with the stamp of “a channel is formed when the water flows” in the below small pond to reflect the business operation philosophy of the enterprise. The materials such as white jade lined marble, copper-coated stainless steel plate as well as the plastic floor tile etc. have conspicuously shown the space qualities. Together with application of the large block of model and wood slices as well as glass, the space looks more connotative and meaningful. The jumping red orange in the exhibition hall is its symbolic color and forms a contrast with the blue light so that the space looks changeable and undulated.

LUMI LEGEND

绿色植物点缀空间。

The green plants decorate the space.

展示厅跳跃的橘红色是企业标志色。

The jumping red orange in the exhibition hall is the symbolic color of the enterprise.

蓝色的灯光现代内敛。

The blue light looks modern and restrained.

大块面的造型简洁明了。

The model in large block looks simple and clear.

国贸嘉和办公楼 Guomao Jiahe Office Building

设计单位:国贸嘉和建筑工程有限公司 / 设计:周浩明 / 参与设计:郭晓阳、李雪霞 / 面积:4500 m^2 / 主要材料:金邦水泥板、免漆饰面板、钢化玻璃、陶条砖、抛光砖 / 坐落地点:苏州 / 工程造价:600万元 / 完工时间:2008年6月 / 摄影:贾立明

Design Unit:JIAHE Construction Engineering Co., Ltd. / Design:Zhou Haoming / Participatory Design:Guo Xiaoyang li Xuexia / Area:4500 m^2 / Main Materials:floor slab、veneer、tempered glass、polished tiles / Location:Suzhou / Project Cost:RMB 6 000 000 / Finish Time:Jun,2008 / Photography:Jia Liming

本建筑设计为T房办公，不管从建筑外观及室内分割都过于对称及呆板。打破这种呆板的毫无激情的室内空间，创造一个全新的具有非凡生命力及创造力的国贸嘉和公司是本案设计的切入点。

室内以“聚”为主题的雕塑，把两幢楼联系起来，为整个小区注入了生命与活力。室内设计手法力求创新与随意，不管是大厅顶面大小不一的无规则秩序的筒灯组合，还是白天、晚上形式阴与阳的变幻对比的无规则发光条，都力求使空间具有耳目一新的视觉冲击。

This building has been designed into a T-shaped office building. No matter whether its visual appearance or the interior division, they are all excessively symmetric and stiff. It is the aim of this design to break through such dull interior space without any enthusiasm and build up a brand-new Guomao Jiahe company with unusual vitality and creativity.

The sculpture subjected as "Gathering" in the indoor has connected two buildings together and brought the life and vigor to the entire region. The indoor design technique strives for the innovation and casualness. No matter from the down lamps of different sizes in an irregular order on the ceiling of the hall, or the irregular luminescent strips in changeable contrast at daytime and night, they try to endow the space with a brand-new and fresh visual impact.

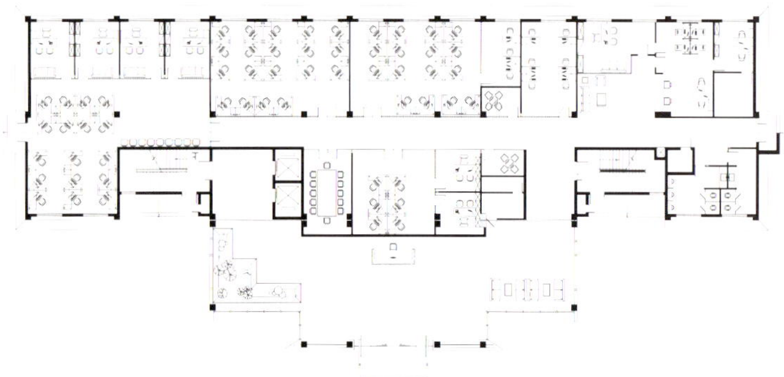

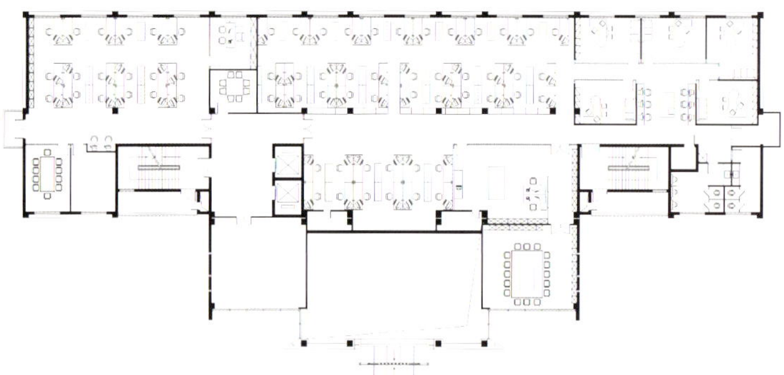

夜色中张扬的现代感。
The prominent modern sense at night.
色彩抢眼的外观充满活力。
The color-conspicuous appearance is full of vitalities.

饰物力求创新，体现非凡激情。

The decorations strive for the innovation and reflect the unusual enthusiasm.

顶面不规则秩序发光条的变幻对比。

The changeable contrast among the luminescent strips in irregular order on top.

黑色竖线条呼应地面白色长短线条。

The black vertical lines echo with the white long and short lines on the ground.

黑白空间隐寓白天和黑夜。

The black and white space implies the daytime and nighttime.

大厅一角。

One corner of the lobby.

黑白色组合带来的强烈视觉冲击。

The strong visual impact brought by the black and white combination.

顶部发光条的条条映射，随意随性。

The reflection of the luminescent strips looks casual and random.

宝兴集团 PROGEN

设计单位:穆哈地设计咨询（上海）有限公司 / 设计:颜呈勋 / 参与设计:陈二琳 / 面积:200 m^2 / 主要材料:钢化玻璃、镜面不锈钢、软包柔力天花 / 坐落地点:上海 / 完工时间:2008年9月 / 摄影:申强

Design Unit:MRT DESIGN / Design:Bill Yen / Participatory Design:Chen Erlin / Area:200 m^2 / Main Materials:tempered glass、mirror finished stainless steel、soft roll / Location:Shanghai / Finish Time: Sep,2008 / Photography:Shen qiang

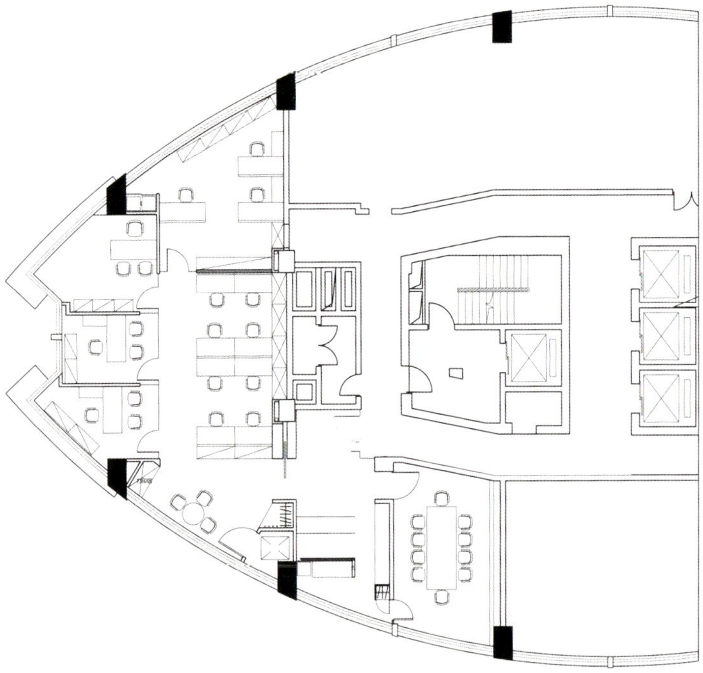

利用材料和设备，打造一个新科技时代的空间。采用封闭与开放式相结合，让区域动线得到更合理的组织。秉承简洁、舒适的理念，合理使用空间，在产品陈列和办公区之间作出恰当分配。天花板的设计加强了视觉识别性，由几片矩形白色柔力天花和石膏板错落分布，给员工一个和谐的工作环境。

接待区为企业形象产品做了陈列柜。由于产品大小不同，做了可移动层板，可根据需要随时更动。入口到背景墙根据LOGO色彩元素红色以及企业形象而定，形成一条装饰带。经理室神秘、优雅。钢化玻璃扮演了极为重要的角色。隔断采用黑色半透明钢化玻璃，结合天花发出的光感，塑造了一个宁静且私密的环境。

the materials and devices are applied to create a space in the era of new technologies. The close and opening divisions are applied together so that the active lines in this area get arranged more reasonably.

The reception area has been made with a display cabinet for the typical products of the enterprise. As the product sizes are different, the removable boards are provided technically so that they can be moved if necessary. A decoration strip from the entrance to the background wall is formed according to the color element of LOGO –red and the image of the enterprise. The manager offices are mysterious and elegant. The toughened glass has played an extremely important role. The partition has applied the black translucent toughened glass to create a quiet and intimate environment together with the light emitted from the ceiling.

This design has taken the philosophy of simplicity and comfortableness. The appropriate arrangement has been made between the product display and office area. The design of the ceiling has strengthened the visual identification. Several rectangular white flexible ceiling boards and gypsum boards have been strewn at random to provide a harmonious working environment for staff.

入口处的企业LOGO。
The Logo of the enterprise at the entrance.
入口到背景墙形成一条装饰带。
It forms a decoration strip from the entrance to the background wall.

ProGen. 宝兴集团
ProGen. Group

办公区天花板由矩形白色天花和石膏板错落分布。

The ceiling in the office area is distributed and strewn at random by the rectangular white ceiling and gypsum board.

根据产品不同大小定做的可移动层板。

The movable boards are customized according to the different sizes of the products.

简洁的设计，封闭与开放相结合。

The simple design, the close and opening divisions are applied together.

通道处的区域动线得到合理的组织。

The active lines in this area of the passage get arranged more reasonably.

红色的局部点缀。

The partial red decoration.

体现企业全新科技时代的背景墙面。

The background wall that reflecting the new technological era of the enterprise.

会议室充满了现代科技的气息。

The conference room is full of modern technological atmosphere.

贝安公司总部 Headquarters of Bam

设计单位:上海米文建筑设计有限公司 / 设计:卢志刚 / 参与设计:李瑞寅、李俊 / 面积:2119 m^2 / 主要材料:钢、玻璃、石材、铝板 / 坐落地点:无锡 / 工程造价:500万元 / 完工时间:2008年11月 / 摄影:刘其华、卢志刚

Design Unit:Shanghai MINAX Architectural Design Co.,Ltd. / Design:Lu Zhigang /Participatory Design: Li Ruiyin、Li jun / Area:2119 m^2 / Main Materials: steel、glass、stone、aluminium panel / Location:Wuxi / Project Cost:RMB 5 000 000 / Finish Time:Nov,2008 / Photography: Liu Qihua、Lu Zhigang

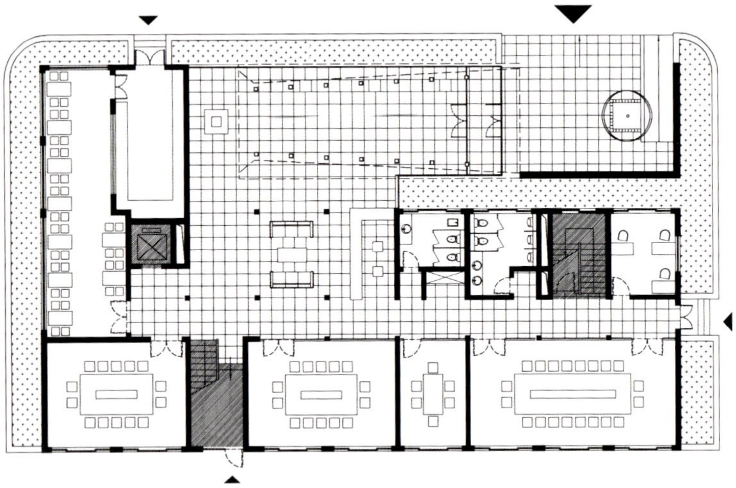

由建筑延续至室内的统一设计，着力于对公共空间秩序的研究与表达，增加办公楼中公共空间的戏剧感与趣味性。不但在横向、竖向的结构与装饰上表达一种秩序感，更在空间的运动中蕴涵这种序列变化所带来的体验，使建筑不仅仅满足于使用的要求，而更多的为使用者带来凌驾于实用功能之上的运动中的诗意。两端悬浮庭院中的运动感受是由心底产生的，镂空的钢梯只是为这样的空间体验提供了一个平台，它可以被视为一个拥有不同高度的看台，舞台就在四周，需要的只是看者调整高度、角度和心情。建筑内部的公共大堂北侧也隐藏着这样一个空间。在通过中部楼梯到达第一个楼梯平台的两侧，竖向的垂直庭院使通高的空间显得狭长而幽静，躲藏在四堵实墙之后的庭院无法一目了然，只有通过在不同的楼层上眺下瞰，才能一窥其真面目。

By the unification design from the construction extends to the interior, we try to research and express on the public spatial order, increase in the dramatization and interesting of the public space in the office building. We try to express a kind of feeling of order from both the structure in thwart and vertical and the decoration, and also the experience in the spatial movement which the change of sequence brings. Thus the building will not only satisfy the requirement of using, but also bring motive poetry above the utilitarian function to the user. Aerosol courtyard in the both sides produces the feeling of movement, and the hollow steel ladder only provides a platform for such spatial experience, it might regard as a stand which has different height, the stage surrounds, people can adjust the altitude, the angle and the mood at will. There is another space like this hides in the north side of the public lobby. Through the central staircase people can arrive at the first terrace, on the both sides, vertical courtyard cause space pass high appears long , narrow and quiet. The courtyard hides after four solid walls, people only look up and down on the different floors, can peep its true colors.

端墙。 End Wall.
大堂上空。 Ceiling Above the Hall.
光的大厅。 Vacant Hall.

BAM

深色墙体内的楼梯。 The stairs inside the dark wall.
二层平台。 Two-story platform.
悬浮的庭院。 Suspension Courtyard.

光孔。 Light hole.
大会议室空间。 Space of large conference room.
小会议室空间。 Space of small meeting room.

BEIJING
NEW YORK
LONDON

LKK室内设计有限公司 LKK Interior Design Co., Ltd.

设计单位:LKK室内设计有限公司 / 设计:林开新 / 面积:150 m² / 主要材料:PVC地板、墙纸、西班牙涂料、木面板 / 坐落地点:福州 / 工程造价:10万元 / 完工时间:2009年1月 / 摄影:吴永长

Design Unit:LKK Interior Design Co., Ltd. / Design:Lin Kaixin / Area:150 m² / Main Materials:PVCfloor、wallpaper、paint、wooden panel / Location:Fuzhou / Project Cost:RMB 100 000 / Finish Time:Jan,2009 / Photography:Wu Yongchang

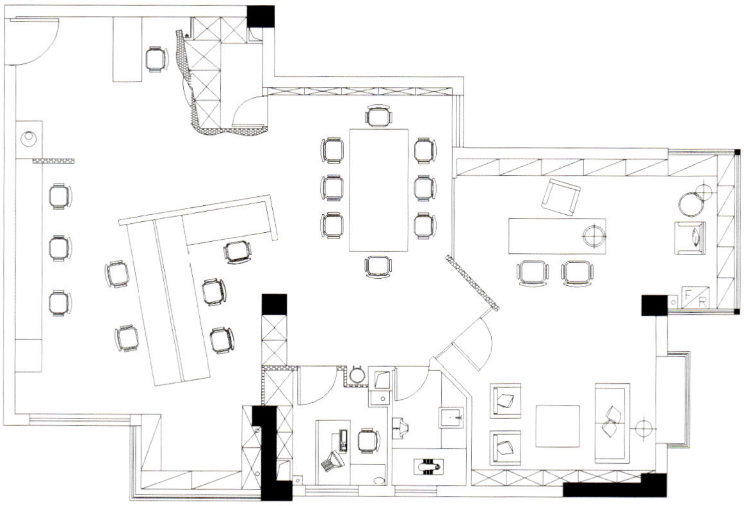

本案是一个室内设计为主营的小型事务所的办公室，设计彰显简洁、时尚，创造高品位的室内空间。空间规划上，工作区域靠窗布置，利用了自然采光和通风。以斜面、曲线有序地组织各功能空间的关系，使界面以自然生长的姿态存在。采用白、灰色为基调，结合少许条纹墙纸、彩色条纹文件盒，使空间简洁而时尚。入口以多层板曲线雕塑为玄关，灵感来源于建筑规划设计上的等高线，把等高线平面立体化，不仅生动了空间，还可在多层板突出处放置书籍及装饰品。玻璃隔断用类似等高线平面的图案呼应玄关，使空间效果统一。主管办公室把书柜艺术化，用方格设计，每个方格都是隐藏的书柜，具有装饰味道的书籍放在显露处给视觉提供美感。

It is a small studio mainly engaged in interior design. It is designed to be simple and fashionable to create an interior space of high taste. In regard to its space planning, the working area is arranged beside the windows with full use of the natural day lighting and ventilation. The inclined surfaces and curves are applied to arrange the relationship between each functional space and make the interface existent in a way of natural growth. White and gray are taken as the main color. In combination with a little stripe wall paper, the space becomes simple and fashionable. The multilayer curve sculpture is applied at the entrance as the front door area. Its inspiration comes from the contour lines in the architectural planning and design. The contour plane is solidified so that not only the space becomes vivid, but also the books and furnishings can be put on the protruded area of the multi-layer board. The glass partition echoes with the front door in a similar contour plane so that the space effects can be uniform. The manager office has made the bookcase artistic. The pane design is applied and every pane is a hidden bookcase. The decorative books are put in the obvious position to provide a visual aesthetic feeling.

入口处以多层板曲线雕塑玄关，生动有趣。
The multilayer curve sculpture is applied at the entrance as the front door area, vivid and interesting.
多层板突出处可放置书籍及装饰品。
The books and furnishings can be put on the protruded area of the multi-layer board.

玻璃增加通透性。

Increase the transparency of the glass.

斜面和曲线有序地组织各功能空间的关系。

The inclined surfaces and curves arrange the relationships among each functional space in order.

玻璃隔断采用类似等高线平面的图案，以呼应玄关。

The glass partition has applied the similar contour plane patterns to echo with the front door.

简约白、灰为基调色。

The simple white and gray are basic colors.

条纹墙纸和彩色条纹文件盒给空间增加变化和色彩。

The striped wall paper and colored striped file boxes have increased the changes and colors of the space.

艺术化的书柜，每个方格设计都是隐藏的书柜。

In the artistic book cabinet, each pane is designed with a hidden bookcase.

纯净时尚的空间。

The pure and fashionable space.

优佰文化办公室 Youbai Culture Office

设计单位:成都优佰文化传播有限公司 / 设计:孙亮 / 面积:90 m^2 / 主要材料:乳胶漆、地板漆、白橡木、玻璃 / 坐落地点:成都 / 工程造价:8万元 / 完工时间:2009年6月 / 摄影:视觉空间商业摄影

Design Unit:You Bai Culture Spreading Co., Ltd. / Design:Sun liang / Area:90 m^2 / Main Materials:emulsion paint、lacquer for floor、white oak、glass / Location:Chengdu / Project Cost:RMB 80 000 / Finish Time: Jun,2009 / Photography:VISION SPACE

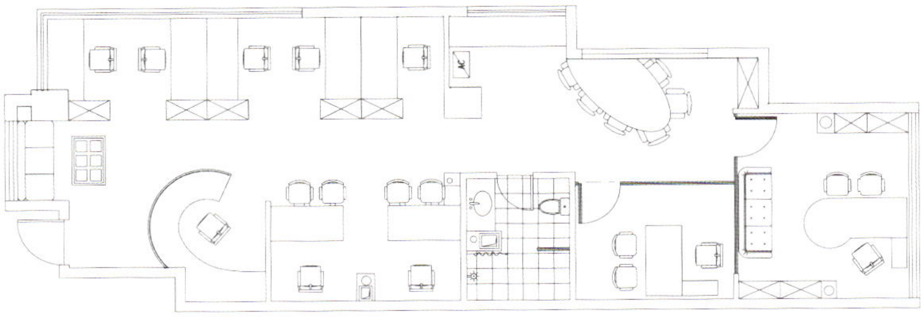

设计师首先对房屋结构做调整，打破原本繁杂的结构，把空间变得开阔且有趣，同时增加了房屋的实用性。透明材质及白色地板漆，使空间明亮、通透；原木色家具及家庭化的布艺窗帘，则是为了保持环境的温馨、舒适；局部绿色玻璃贴纸和植物的点缀，既符合公司整体形象色彩，又打破了办公空间的沉闷。风格简洁、流畅，通过简单的手法表现丰富的内涵。值得一提的是，整个方案所使用的都是常见并且廉价的装饰材料，体现了公司一贯的创作原则：设计的精髓在于思想，而不是材料或造型堆砌。

The designer first has adjusted the room structure and broken through the original complicated structure so that the space becomes wide and interesting and the practicality of the room is also increased. The transparent materials and white floor paint make the space bright and transparent; the original log color furniture and the family cloth curtains are applied to maintain the warmness and comfortableness of the environment; the spot decoration of green glass paster and plants in partial areas both match the colors of the company' s entire image and also break through the oppression of the office space. The style is simple and smooth and the rich meanings are expressed through the simple techniques. What is worth mentioning is that the common and cheap decoration materials are applied in the whole scheme, which has reflected the consistent creation principle: the essence of a design lies in the ideals instead of the piling of materials or moulding.

调整后宽敞通透的空间。
The spacious and transparent space after adjustment.
采用简单常见的材料表现温馨舒适。
The simple and common materials are applied to show the warmness and comfort.
前台接待处。
The reception area.

优佰文化

透明材质及白色地板漆使空间明亮。

The transparent materials and white floor board paint brighten the space.

木条块组合起的吊顶贯穿整体。

The ceiling assembled by the wood stripes runs through the whole.

玻璃上的绿色贴纸贴近自然。

The green adhesive papers on the glass are close to the nature.

昆山振华建设集团办公楼

Kunshan Zhenhua Construction Group Office Block

设计单位:金螳螂第一设计分院一所 / 设计:王祎华、钱文宇、糜佳、刘岩、王启霖 / 面积:10500 m^2 / 主要材料:灰色机刨花岗岩、灰木纹石、金属天花、金属防火板、透光软膜结构 / 坐落地点:昆山 / 工程造价:1100万元 / 完工时间:2009年4月 / 摄影:王祎华

Design Unit:Gold Mantis Designing Institute / Design:Wang Yihua、Qian Wenyu、Mi jia、Liu yan、Wang Qilin / Area:10500 m^2 / Main Materials:granite、stone、fire resisting shield、pia mater / Location:Kunshan / Project Cost:RMB 11 000 000 / Finish Time:Apr,2009 / Photography:Wang Yihua

业主期望新旧建筑连接组合后，满足集团办公细化功能要求，打造行业龙头建筑企业的办公空间形象。设计首先就业主的使用功能进行细化研讨，明确室内设计项目本身作为一个产品，所要服务的目标市场定位。同时总协调建筑、幕墙、机电、照明、给排水、暖通、弱电的无缝对接，精确测量建筑各层实际状况。空间形式中正、对称，以较新的材料和工艺显示作为一个龙头建设单位的沉稳而又不失创新的企业文化。

The clients hope the combination of the new block and the original building can satisfy the detailed function business demand of the company group, and to erect an office space which is suit for a leading construction company of the industry. First, the designers discuss with the clients about the detailed user functions, making sure the the market position of the service target when regarding the interior design itself as a product. At the same time, we coordinate with different teams such as construction, cuitainwall, mechanical and electrical, llumination, water supply and drainage, HVAC, and light current. We accurately measure each level' s actual state. The space style is upright and symmetrical. Relatively new material and craftsmanship express the corporate culture of a leading construction company.

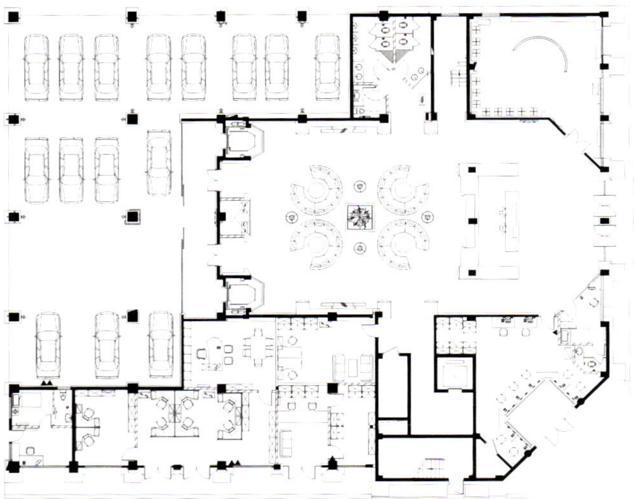

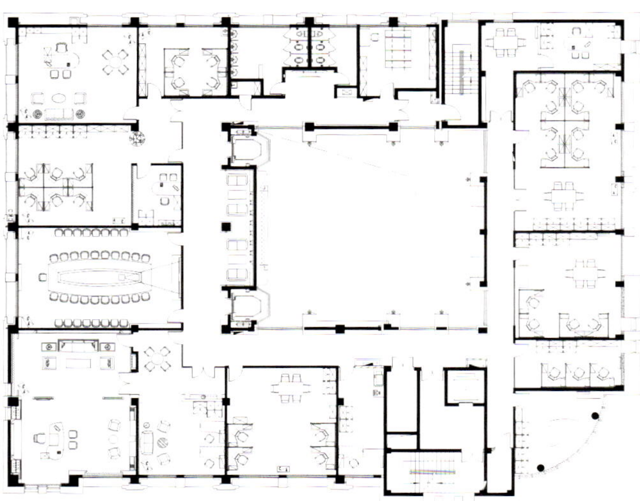

中正、对称、大气的空间形式。

The fair, symmetric and grand space form.

康恩贝大厦 Conba Mansion

设计单位:杭州典尚建筑装饰设计有限公司 / 设计主持:陈耀光 / 参与设计:胡昕、朱玉萍、刘伟 / 面积:1250 m^2 / 主要材料:大理石、木饰面、磨砂玻璃、地毯 / 坐落地点:杭州 / 工程造价:250万元 / 完工时间:2008年6月 / 摄影:陈乙

Design Unit:Hangzhou Dianshang Building Decoration Design Co., Ltd / Principal Design:Chen Yaoguang /Participatory Design: Hu xin、Zhu Yuping、Liu wei / Area:1250 m^2 / Main Materials: marble、wood veneer、frosted glass、carpet / Location:Hangzhou / Project Cost:RMB 2 500 000 / Finish Time:Jun,2008 / Photography:Chen yi

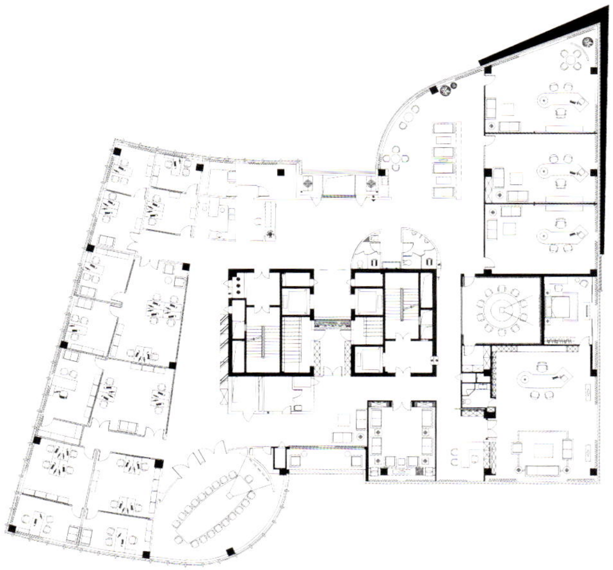

当代企业商务写字楼室内设计的核心工作除了对区域、功能划分和现代办公设施完善的配备以外，公共部分的空间气质是设计的关键。它可以展示企业的性质，可以体现企业的文化，可以传递企业的精神面貌。康恩贝企业最初从养蜂起家，我们以蜂窝的元素分别用于空间的界面上，整体而不孤立，传承又不失现代感。灯光、形态还有色彩等元素的有效组合，充分体现一种高科技、高智能、高效益的具有国际商务办公空间的精神气质。主材为木饰面和大理石，现代而不失温馨，商务而不失亲和。

In addition that the key work of interior design for the contemporary business office buildings is to divide the areas and functions as well as provide the complete modern office facilities, the space atmosphere of the public part is the key of the design. It can show the characteristics of the enterprise, reflect its culture and deliver its spiritual outlook. Conba enterprise built up from beekeeping. We have applied the honeycomb element to the space interfaces separately. It looks integral, but not isolated, continued but without loss of modernity. The effective combination of the light, forms as well as the colors has fully revealed a hi-tech, high-intelligent, high-efficient spiritual atmosphere with international business office spaces. The main materials are the wood veneers and marble, modern but warm, commercial but friendly.

公共休息区。
Public Rest Area.
接待大厅。
The Reception Hall.

康恩贝集团

门厅主立面。
Main Facade of Entrance Hall.
门厅主立面设备暗门。
Concealed Equipment Door at Main Facade of Entrance Hall.
企业文化展示长廊。
Culture Exhibition Corridor of Enterprise.
公共盥洗室。
Public Washroom.

公益基金会办公室 Office of Public Welfare Fund

设计单位:BDM Design / 设计:王宏、周捷 / 面积:300 m² / 主要材料:花梨木、金砖、丝绸墙纸、酸洗玻璃、地毯 / 坐落地点:北京 / 工程造价:200万元 / 完工时间:2008年6月 / 摄影:孙翔宇

Design Unit:BDM Design / Design:Wang hong、Zhou jie / Area:300 m² / Main Materials:rosewood、gold brick、wallpaper、glass、carpet /Location:Beijing / Project Cost:RMB 2 000 000 / Finish Time:Jun,2008 / Photography:Sun Xiangyu

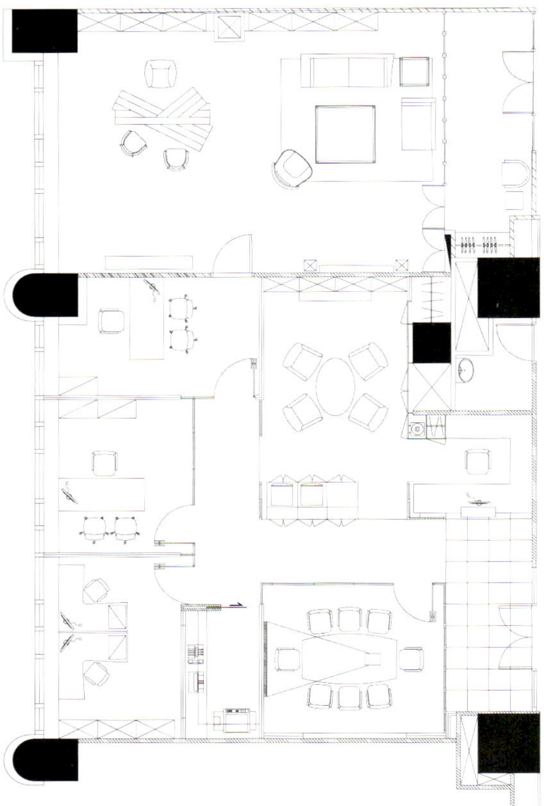

这是一间非营利性公益基金会的办公室，主要从事资助对我国文化、教育、经济等各方面有益的研究及项目，而非一般的商业活动，因此这里更像一个书房及会友的空间。功能决定属性，总体来说是一次古典中式与现代化西式元素的和平共处，包含了既融合又充满碰撞的设计理念和处理手法。

中式古典屏风从入口处便告诉人们这里的文化特质，不同的组合及材质变化使整个空间灵动统一；灰色的地面和方块地毯既满足了不同区域的功能需求，又从色彩上达成高度统一。除了红色和灰色这两种颇具中国韵味的色彩外，便是大面积的留白和点缀其间的黑色，但这里的白色绝非平淡的退让，而是由包括意大利暗纹丝绸壁纸、哑光烤漆墙板、人造石柜体台面和酸洗玻璃等构成丰富细腻的变化。

其间的活动家具和装饰陈设使空间更具活力，这里有Philippe Strack的白色沙发和Friz Hansen的红色蛋椅，他们的下面是一块手工编织的羊毛藏毯；Cassina的书桌椅配上了Herman Miller的叶子台灯；在装有B&O的极简电视和音响的同一面墙上是纯中式的花梨条案和落地灯架；在休息区有天鹅椅、花梨木屏风，白色人造石矮柜上的青花瓷盆中有几条红色金鱼在轻快地游动。

It is an office of a non-profit public welfare fund. The function decides the property. Generally speaking, it is a harmonious coexistence of classic Chinese and modern western elements, comprising both the integrated and conflicting design philosophy and processing techniques.

The Chinese style classic screen at the entrance tells us the cultural characteristics here. The different combinations and material changes have made the whole space bright and unified; in addition to the two kinds of quite charming colors – red and gray, the large area of blanks and the dotted black also have constituted the wonderful and exquisite changes.

The furniture and displays among them have vivified the space much more. Here is the white sofa, red egg chair and hand woven wool Tibetan carpets; the desk and chairs are arranged with the leaf-shaped table light while the wall is hung with the pure Chinese-style rosewood long narrow table and the floor lamp rack; there are the swan-shaped chairs, rosewood folding screens. There are also several golden fish swimming briskly in blue and white porcelain pot on the low white artificial stone cabinet.

古典中式与西式元素在此和谐相处。
The classic Chinese and modern western elements coexist harmoniously.
中国红和大面积白色的交互点缀。
The mutual decoration of Chinese red and large area of white.

家具陈设富有浓郁的文化气息，正契合业主的性质。

The furniture display has a strong cultural charm and just matches the owner's characteristics.

纯中式落地灯架后是暗纹丝绸壁纸，细腻婉约。

Behind the pure Chinese-style floor lamp rack is dim patterned silk wall paper, exquisite and graceful.

中式屏风和现代玻璃的碰撞组合。

The collision and combination of Chinese screen and modern glass.

更像书房和会友的空间。

A space more like a study room and for gathering.

大地旅行社办公室 Dadi Travel Agency

设计单位:瀚禹・印加空间设计机构 / 设计:吴君华 / 参与设计:白晓晓 / 面积:150 m^2 / 主要材料:玻璃、大理石、钢、复合地板、肌理漆、抛光砖 / 坐落地点:宁波 / 工程造价:8万元 / 完工时间:2009年6月 / 摄影:张学泉

Design Unit:Hanyu・Yinjia Space Design / Design:Wu Junhua / Participatory Design:Bai Xiaoxiao / Area:150 m^2 / Main Materials:glass、marble、steel、laminate flooring、paint、polished tile / Location: Ningbo / Project Cost:RMB 80 000 / Finish Time:Jun,2009 / Photography:Zhang Xuequan

设计师采用简洁明快的装饰手法，在保持功能与美观的条件下，完成个性化的创造与表现。空间上点、线、面结合，主体以白色为基调，从平面、立体转换到家具和饰品，再加入材质间的粗细对比，演变出极具立体感的经典格调。为避免黑白色调的单调，设计师运用灰色作为过渡，深色地毯与浅色地砖的运用，令空间层次丰富。空间的划分采用玻璃，不仅让视觉达到最大程度的延伸，又让空间显得通透。细节处理上加入柔和元素，彰显独特魅力。

The designer has applied a simple and bright decoration technique to accomplish the unique creation and expression while maintaining the functions and beauty. The dots, lines and surfaces are combined in this space while the main structure has taken white as its main tone. A classic style of extreme third dimensional sense has been depicted through the change from plan and three-dimension to the furniture and furnishing in addition to the contrast between the thickness and thinness of the materials. To avoid the monotony of black and white colors, the designer has applied gray as a transition. The application of the dark carpet and light ground tiles has enriched the space levels. The glass is used for the division between the space, which not only has reached the maximum extension visually, but also has made the space more transparent. The arrangement on the details together with the soft elements has shown its unique charms.

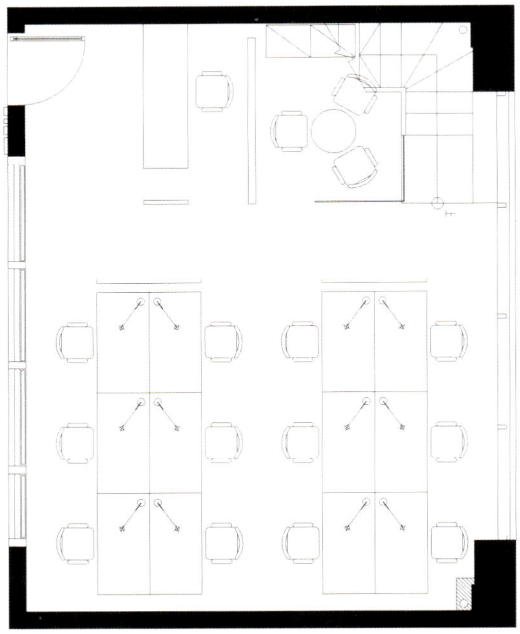

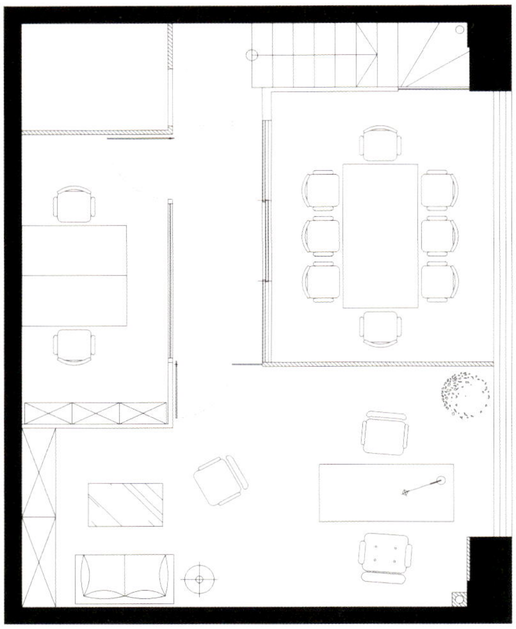

黑白色调简洁明了。

The black and white colors look simple and clear.

空间上点、线、面的组合具有立体感。

The combination of dots, lines and surfaces in the space looks three-dimensional sense.

楼梯的简洁处理。

The concise arrangement of the staircase.

通透的办公空间。

The transparent office space.

深色地毯的运用使空间层次丰富。

The application of dark colored carpet enriches the space levels.

华劲集团股份有限公司 Hwagain Group Incorporated Company

设计单位:深圳市图人设计有限公司 / 设计:周际 / 参与设计:辛志为、杨斯、张志成、严芝敏 / 面积:3000 m² 主要材料:木化石、瑞士梨木、白橡木、微孔铝板、乐宝方块地毯、雾化玻璃、亚克力板 / 坐落地点:南宁 / 工程造价:600万元 / 完工时间:2009年2月 / 摄影:吴永长

Design Unit:TRD / Design:Zhou ji / Participatory Design:Xin Zhiwei、Yang si、Zhang Zhicheng、Yan Zhimin / Area:3000 m² / Main Materials:petrified wood、swiss pear、white oak、aluminium plate、carpet、glass、acrylic board / Location:Nanning / Project Cost:RMB 6 000 000 / Finish Time:Feb,2009 / Photography:Wu Yongchang

由于集团部门多，人数多，考虑到空间的自然采光和通风，设计师将其规划整合，使各功能之间流线通畅，疏而不散，密而不封。

在接待大厅的设计中，打破常规地将接待台设在入口的右侧，使室外的自然光通过休息区进入室内，在大厅与高管办公区之间用一水景Logo形象墙划分，将各功能区沿窗而设，充分利用室外景观和自然采光。22层董事长办公区的布局，借鉴了中国古典园林中的造园手法。通过对空间进行分割与联系，产生一种深远的意境，整个空间既有分割又有连通。秘书办精简纯净，折合垂落的木化石造型呼应了下方的白橡木接待台，并与墙面浑然一体。漫射光和反射光的照明给人以舒适感。空间通过Logo形象墙镂空处与叠层石材浅水池相连，形成景观等候区。董事长会客室色彩凝重、温和、舒适，具有独特的文化韵味。设计注重人与自然的亲切交流，使室内极具观赏性。方形雾化玻璃吊灯呼应围合的天花，通往会议室的门设计成落地的旋转门，灵活使用，开启时空间通透，关闭时空间规整。靠窗处设梨木框套雾化玻璃屏风，在其圆形镂空处摆放工艺品。透过屏风和植物望向窗外，室外的景色则被巧妙地引入室内，这种借鉴苏州园林里常用的虚实对比的手法，颇有“镜中风景入画屏”的意境。

The design of the reception hall has unconventionally arranged the reception desk on the right so that the natural light can enter the indoor. The waterscape logo wall is applied to divide the hall and senior management offices. Each functional area is planned along the windows in order to make full use of the outdoor scenery and day lighting. The arrangement of the office area has referred to the gardening technique of the classic gardens. A profound artistic atmosphere is created through the division and connection. The secretary office is compacted, simple and pure. The downward wood fossil sculpture echoes with the below white oak reception table and integrated into one with the wall. The illumination of the diffused and reflected light provides a comfortable feeling. The space is connected through the hollowed-out area in the LOGO image wall and the shallow water pond of the piled stone materials to form a scenic waiting area. The visitor reception area has mild and comfortable colors and emphasizes on the communication between human and nature so that the indoor is admirable. The square frosted glass pendant lamp echoes with the ceiling while the door to the conference room is designed as a floored rotating door so that the space is transparent while opening and clear and net while closing. The frosted glass screen framed with the pear wood is designed beside the window. Watching outside the window through the screen, the outdoor scene is tactfully introduced into the indoor, which quite has an artistic atmosphere of "Scenery in the mirror enters the painting screen".

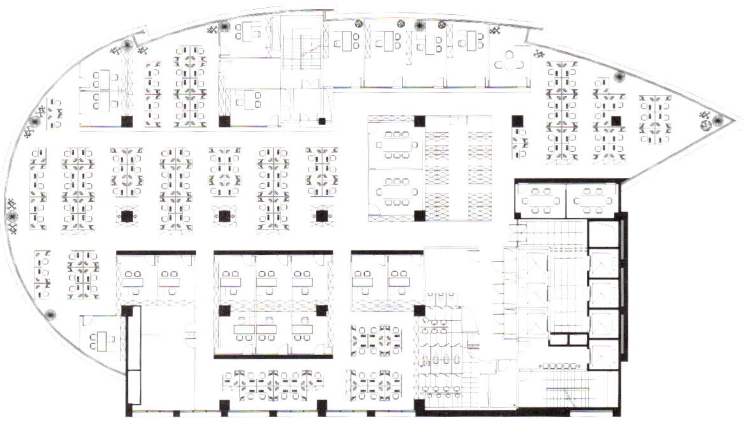

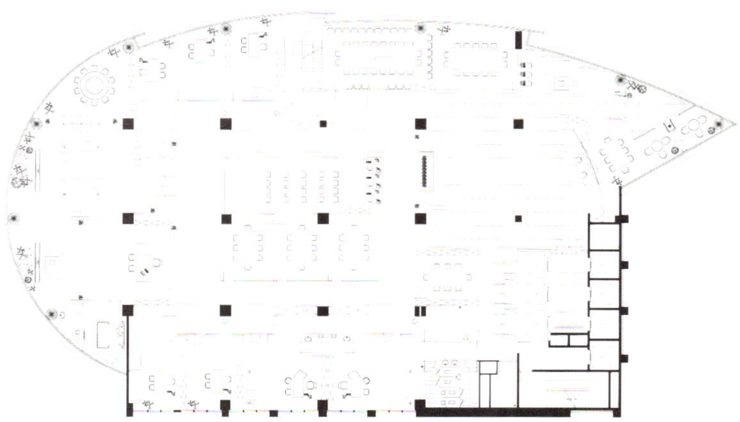

空间精简纯净。
The space is concise and pure.
宽敞明亮的室内。
The spacious and bright indoor.

极具观赏性的装饰。
The decoration extremely worth viewing.
办公区。
Office Area.
洽谈室。
Discussion Room.
室外景色引入室内。
The outdoor sight is introduced into the indoor.

空间既有分割又有连通。
The space is both divided and connected.
落地旋转门使用灵活。
The revolving door can be used flexibly.
方形雾化玻璃吊灯呼应围合的天花。
The square frosted glass pendant lamp echoes with the enclosed ceiling.
色彩凝重、温和、舒适。
The colors look dignified, mild and comfortable.

阿里巴巴滨江总部办公大楼 Office Building for Alibaba Binjiang Headquarters

设计单位:内建筑设计事务所 / 面积:120000 m^2 / 主要材料:黑水晶石材、地胶板、橡木饰面、丝网印玻璃 / 坐落地点:杭州 / 工程造价:6000万元 / 完工时间:2009年8月 / 摄影:潘杰

Design Unit:Interior Architects Design / Area:120000 m^2 / Main Materials:morion stone、rubber plates oak surface、glass / Location:Hangzhou / Project Cost:RMB 60 000 000 / Finish Time:Aug,2009 / Photography:Pan jie

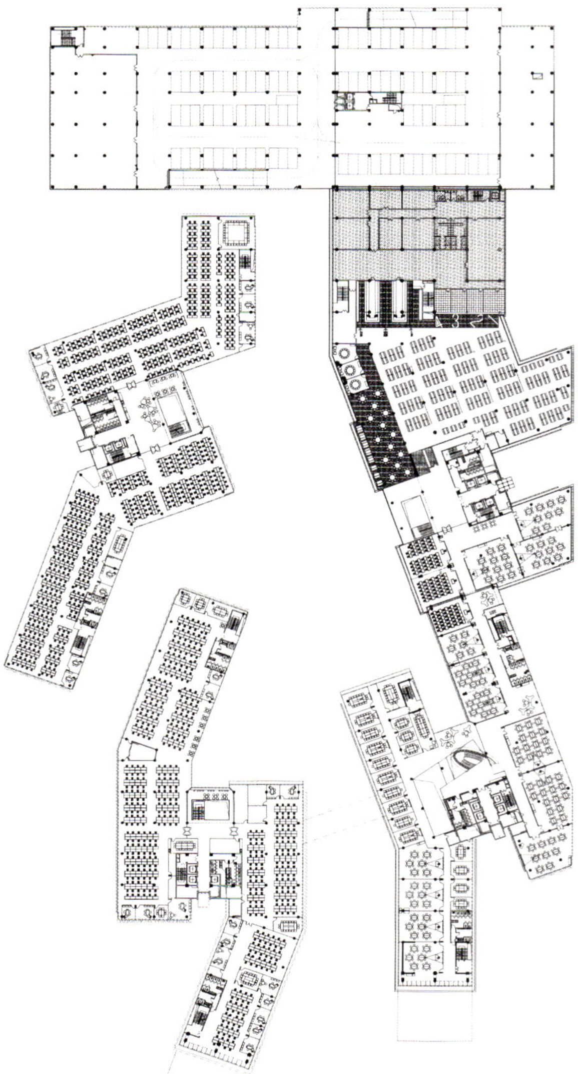

对阿里巴巴来说，此次的一大变化是集合。原本分散在不同写字楼中的公司群体集中进驻这个建筑综合体，与之相对应的也是功能需求上的整合。邮局、银行、餐厅、书店、健身中心、咖啡馆等新鲜生活化因子的加入，打破单调的办公职能。阿里巴巴的标识字样及标志性的橘黄色随处可见。

一号楼一层为公司主入口，以总接待台为界点，左右两翼分别安排等候与会谈不同的功能区域。设计以时间为主线，在西侧空间中把休息等候区与阿里巴巴历史文化展示相交织。两块从地面上翻的多媒体板形成特殊空间和轴线。黑色地面上以银色细线勾勒出斜向路径，路径上嵌以阿里巴巴的英文与公司大事记年份的字母数字组合。

为达到以最少门禁进行最有效控制，保障内部交通顺畅的目的，设计在一层及二层公共区域内安排了大量的会客空间，尽量避免上层办公区域中不必要的外界干扰，维持高效的办公秩序。由玻璃和打孔铝板围合出供2~3人使用的小型六角形蜂窝状的洽谈室，几个相连形成一组，在接待台右侧沿折线展开。

与硬朗的建筑外观形成对比的是室内大量出现的圆造型以及有着柔和线条的家具。圆孔镂空木板、孔洞、白色张拉膜等设计都以弧度来消解拘泥，铺垫出柔软的松弛。

二楼以上的大办公区设计以最便捷的方式加强内部沟通与联系，扩大组群社交范围。大开间格局，不设隔断。四至五层中庭还以滑梯串联上下层空间，趣味设计激发创意动能产生。

设计沿用以往的低技策略，坚持低成本、大创意的设计路线。用廉价材料作创意设计，以快乐色彩补充材质缺点。

For Alibaba, this great change is a collection. The new living elements such as the post office, bank, restaurant, bookstore, fitness center, cafe etc. are joined in to break through the monotonous office function. The logo of Alibaba and the symbolic bisque is visible everywhere.

The first floor of building #1 is the main entrance of the company with general reception desk as the interface, on the left and right sides of which are arranged with the different functional areas for waiting and conversion. The black floor is outlined by the fine silver line with an inclined path, on which are inlayed with the combination of the characters and figures – English name of Alibaba and the years of its memorabilia. The atrium from the fourth to fifth floor has also applied a slide to series connect the upper and lower spaces. The interesting design inspires the generation of initiatives. To achieve the most effective control with minimum access controls and ensure the internal smooth transportation, the designer has arranged a great number of reception spaces in the public areas of the first and second floors in order to best avoid the unnecessary external interferences to the office areas in the upper floors and maintain the high-efficient office system.

What forms a contrast with the hard architectural appearance is the emergence of round shapes and the furniture with soft lines in great number indoors. The round-hole hollowed-out wood boards, holes, white tension films etc are designed in arc to dissolve the constraints and set the soft looseness. The design has continued the previous strategy of low cost and high technology and persists in the design philosophy of low-cost and great creativity. The cheap materials are used for the creative design and the pleasant materials are applied to supplement the disadvantages of the materials.

总接待台。
Reception Desk.

Alibaba.com

不规则造型的家具、隔断。
Irregularly shaped furniture and partition walls.
公共区域内安排了大量的会客空间。
A lot of meeting zones are arranged in the public area.
两块从地面上翻的多媒体板形成特殊空间和轴线，如缎带般起伏舞动。
Two pieces of multimedia boards turning up from the floor form the special space and axes, waving like a satin ribbon.

Alibaba
2009

由白色张拉膜做出球体裹住的楼梯。
White tensioning film forms the stairs covered by a sphere.
白色张拉膜。
Upward view of the white tensioning film.
楼梯外观。
Appearance of the stairs.
沿着地面橘黄色的城市大道可以快速便捷地进入不同的工作区域。
Difference working zones are accessible conveniently along the city avenue of yellow ground.

服务台。
Service desk.
别致的书架。
Unique book shelves.
员工餐厅。
Employee's dining hall.

ook

上海中心 Shanghai tower

设计单位:上海米文建筑设计有限公司 / 设计:卢志刚 / 参与设计:陈斌鑫、李俊、李瑞寅、袁波 / 面积:2338m^2 / 主要材料:钢、玻璃、石材、铝板、涂料 / 坐落地点:上海 / 工程造价:300万元 / 完工时间:2008年12月 / 摄影:刘其华、卢志刚

Design Unit:Shanghai MINAX Architectural Design Co.,Ltd. / Design:Lu Zhigang /Participatory Design: Chen Binxin、Li jun、Li Ruiyin、Yuan bo / Area:2338m^2 / Main Materials: steel、glass、stone、aluminium panel、coating / Location:Shanghai / Project Cost:RMB 3 000 000 / Finish Time:Dec,2008 / Photography: Liu Qihua、Lu Zhigang

如何让一个普通寻常的空间，在外部条件的强势介入下产生积极的变化，从而焕发出新的戏剧化表现呢？此时，对于新空间的期待油然而生。新体量适时的介入，改变了原来建筑的组织秩序。一个矩形的新加建门厅，借用了笛卡尔的透视体系原理，通过围合界面的线性调整，通过夸大透视的收缩效果，将空间线性拉长，并通过序列的钢柱，强化这种进入感，并将视线聚焦至巨大的展示模型，让建筑的进入过程，充满仪式感。建筑空间通过双向延伸，使内部空间因为一种外部的介入而产生碰撞和变化。竖向序列也由原本平直的均一，转化为尺度的波状起伏。黄色和蓝色确定为建筑空间的基本色，高对比产生的强烈反差，透过石材、墙面和灯光表现出来，而高纯度的蓝色成为整个空间的灵魂。

How to make an common space have the positive change, which also radiates new dramatic performance, under the strong intervention of external condition. At this moment, the expectation to the new space rises of itself. The intervention of the new dimension timely, will change organization order of the original construct. A rectangular hall adds to build, using Cartesian perspective system principle, adjusts the round line of interface, exaggerates the contractive effect of perspective, stretch the linearity of the space. And through the sequence of steel columns, make the huge display model focus, strengthen the sense of entering into the construction, make the process like the ceremony . The construction space has bidirectional extends, causes the inner space to have the collision and change because of one kind of exterior involvement. The straight vertical sequence also becomes to undulate. Yellow and blue are the primary color of the construction space, the intense contrast that high contrast produces, shows out by the stone material, wall surface and lamp light , and the high-purity blue becomes the soul of the entire space.

上海中心外观。

The outward appearance of the Shanghai Tower.

门厅与主体的接缝。

The joints of the main hall entrance and the body.

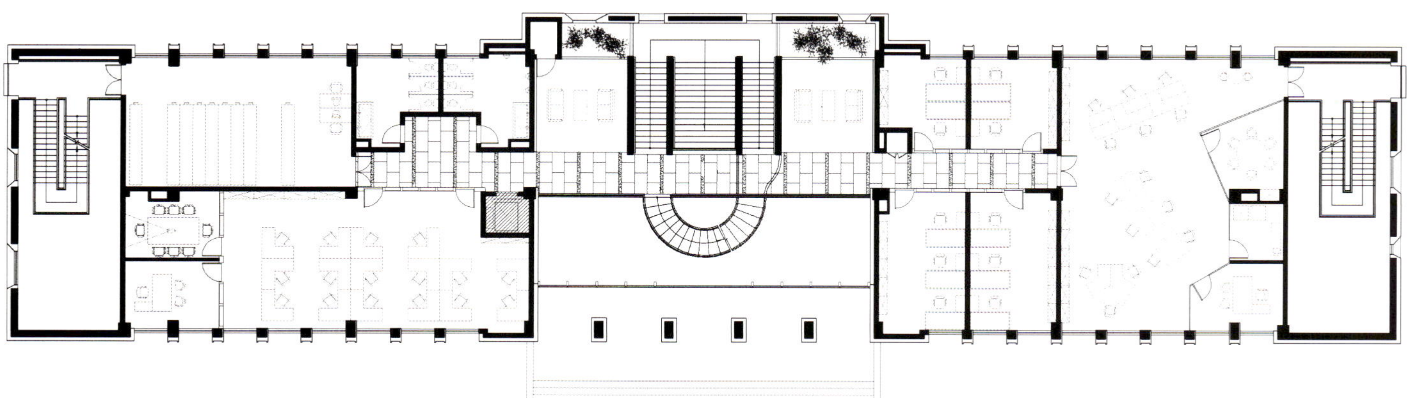

上海中心大厦建设发展有限公司

扁平腔的接待处。
Flat cavity reception.
夜晚室内颜色的不同表情。
Different colors in the night room represent different expressions.
序列的廊柱门厅和下降的顶棚加强了空间的透视感。
Sequence pillars hall and low roof intensify a sense of perspective of the space.
新增加的钢楼梯。
A new added steel stairs.
简单材料形成的对比。
A contrast formed by the simple materials.
不同方向视觉焦点的上海中心模型。
The visual focus to the mock-up of Shanghai Tower from different directions.

山东广电中心 Shandong Broadcasting and Television Center

设计单位:中国建筑设计研究院室内设计研究所 / 设计:张晔 / 面积:100000 m² / 主要材料:石材、印刷玻璃 / 坐落地点:济南 / 完工时间:2009年7月 / 摄影:顾建英

Design Unit:China Architecture Design and Research Group Interior Design Institute / Design:Zhang ye / Area:100000 m² / Main Materials:stone、printed glass / Location:Jinan / Finish Time:Jul,2009 / Photography:Gu Jianying

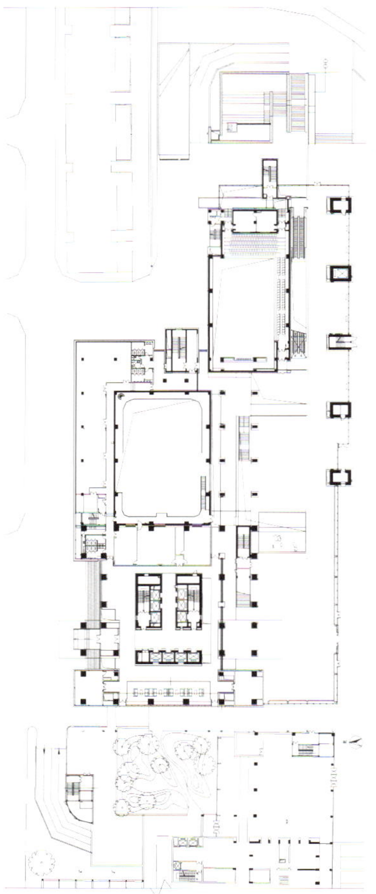

山东广电中心建筑与室内是一个完整的空间设计作品。室内设计延续建筑界面，完成内部空间，在浑然一体的背景下反映山东当地文化、地域特点又带有媒体特征的主题节点，如天书长卷、大屏幕、地下展厅等；在空间中利用现代技术手段营造“媒体之城”空间氛围；大堂内部高19m，两侧超尺度的封闭墙面，置身在其中感受到一种震撼；在大厅中抬头仰望，一卷长约百米的天书漂浮在空中。书文是王羲之的“兰亭序”。这卷“天书”不仅内部很巧妙地暗藏大厅的屋面排水，同时也减少了整个大厅的玻璃屋顶面积，在节能方面发挥着巨大的作用。夜晚，“天书”可以为大厅提供舒适的照明；大堂的一层设一方水池，叠水缓缓漫入地下一层展厅，池中蚀刻遒劲豪放的书法文字，源于泰山的“经石峪”，显示出一种令人肃然起敬、超乎寻常的威势。文字被流水冲刷着，隐喻齐鲁文化的源远流长。山东广电从建筑到室内空间的设计根植于本土文化的基础上，发掘地域文化的精髓，寻找与现代文化的嫁接点，寻找现代设计和地域文化的相融共生。

The architecture and interior of Shandong Broadcasting and Interior are an integral space design work. The interior design continues the architectural interface to complete the interior space and reflect the local cultures in Shandong in a perfectly integrated background. The modern technologic technique is applied in the space to create a space atmosphere of “city of media”; looking up to the hall, an approximately 100m long heavenly book is floating in the sky. The book is “Prologue at Orchid Pavilion” written by a famous Chinese calligrapher Xizhi Wang. This roll of “heavenly book” not only has skillfully hided the roof water drainage system of the hall inside, but also has reduced the glass roof area of the whole hall, which has take great effects on energy saving. At night, the “heavenly book” can provide comfortable lighting to the hall; the first floor of the hall is arranged with a water pool, where the overlapping water is overflowing into the exhibition hall in the underground. The powerful and unconstrained handwriting words are etched in the pool, which comes from “Jing Shi Yu” of Taishan Mountain. The design from the architectural space to interior space in Shandong Broadcasting and Television Center is based on the local cultures to explore the essences of the regional cultures and seek the integration and coexistence of the modern design and regional cultures.

室内设计延续建筑界面，完成内部空间。
The indoor design for the internal space is an extension of the architectural interface.
在大厅中抬头仰望，一卷长约百米的天书漂浮在空中。
A book of about a hundred meter long is floating in the sky when looking up in the hall.
夜晚“天书”可以为大厅提供舒适的照明。
The "sky book" can supply comfortable illuminating for the hall.

灵山梵宫 Lingshan Vatican Palace

设计单位:无锡灵山实业责任有限公司 / 设计:陆嵘 / 参与设计:蔡鑫、鱼晓亮、翁芝青 / 面积:75000m² / 主要材料:石材、木雕、壁纸 / 坐落地点:无锡 / 工程造价:7亿元 / 完工时间:2008年11月 / 摄影: 文宗博、潘宇峰

Design Unit:Wuxi Lingshan Industry Co., Ltd. / Design:Lu rong / Participatory Design:Cai xin、Yu Xiaoliang、Weng Zhiqing / Area:75000m² / Main Materials:stone、wood carving、wallpaper / Location: Wuxi / Project Cost:RMB 700 000 000 / Finish Time:Nov,2008 / Photography:Wen Zongbo、Pan Yufeng

灵山梵宫是继灵山大佛、灵山圣境佛教文化园之后，在无锡灵山圣地启动的又一项令人瞩目的工程，是世界佛教论坛的会址，同时也是灵山圣地的又一胜景，无论是建筑还是室内空间的艺术表达都给来访者留下深刻的印象。

整体梵宫由塔厅、廊厅及圣坛三大区域组成。塔厅空间高耸，廊厅秩序感强烈，圣坛气氛神秘，室内空间尤其注重造型和色彩的运用，浓烈的宗教色彩、震撼的气势，带给人们无限的遐想，达到洗涤心境、领悟人生的境界。

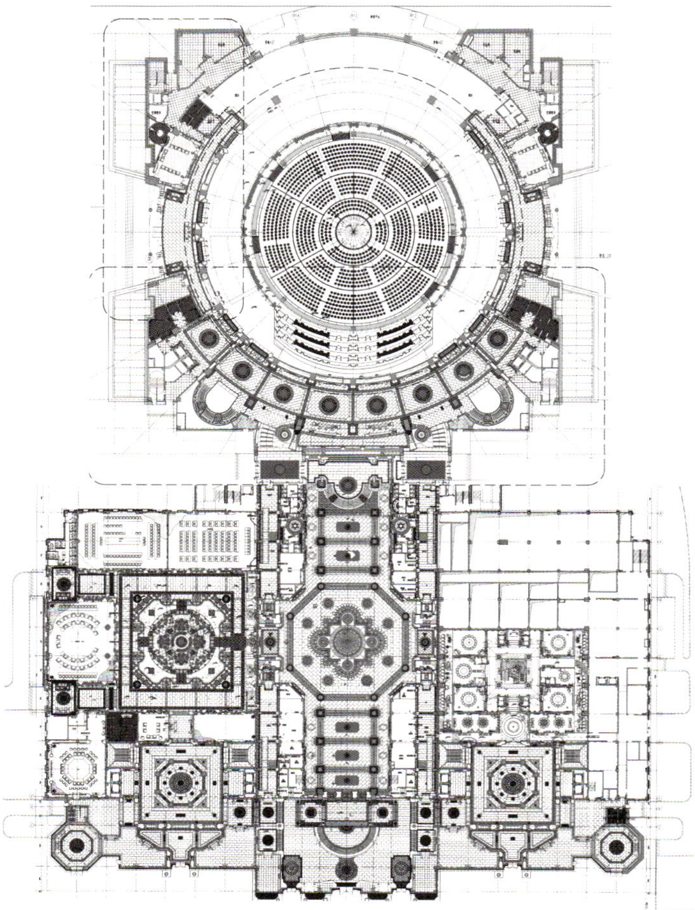

Following the Lingshan Grand Buddha and Lingshan Buddhist Theme park, Lingshan Vatican Palace is another impressive project started in the sacred ground in Lingshan, Wuxi. After its completion it will become the venue of the World Buddhist Forum and also another attraction in the Holy land in Lingshan. Both its building and the artistic expression of its interior space will leave a deep impression on visitors.

The whole Vatican Palace is constituted by the Tower Hall, Gallery Hall and the Altar. The high-rise space in the tower hall and orderly layout of the corridor make the Altar spreading an atmosphere of mystery. The interior spaces are particularly emphasized by using the shape and color. Its intense religious atmospheres and magnificent shape will give people such a endless reverie that wash their mood and arrive the realm of life.

巧夺天工的细部。
The detail with superb workmanship.
金碧辉煌的大厅。
The resplendent and magnificent hall.

精雕细琢的雄伟门廊。
The delicately and finely carved majestic entrance hall.

石材与木材的亲密对话。
The intimate dialogue between the stone and wood materials.
巨型石柱足足有三层楼高。
The giant stone column is three-story high fully.
飞天雕刻惟妙惟肖。
The sculptures of flying fairies are absolutely lifelike.
华丽得可以媲美西方极乐。
It is so magnificent so that it almost can be compared with the western seventh heaven.

圆满和乐的美好祝愿。
The satisfactory, happy and harmonious good wish.
气势恢宏的大空间。
The large gorgeous and magnificent space.

星湾学校 Xingwan School

设计单位:国贸嘉和建筑工程有限公司 / 设计:王严平 / 参与设计:李龙龙、马玲 / 面积:5320 m² / 主要材料:抛光砖、成品玻璃隔断、地毯、硬包 / 坐落地点:苏州 / 工程造价:518万元 / 完工时间:2008年7月 / 摄影:贾立明

Design Unit:Jiahe Construction Engineering Co., Ltd. / Design:Wang Yanping / Participatory Design: Li Longlong Ma ling / Area:5320 m² / Main Materials:polished tiles、glazed wall、carpet、hard roll / Location:Suzhou / Project Cost:RMB 5 180 000 / Finish Time:Jul,2008 / Photography:Jia Liming

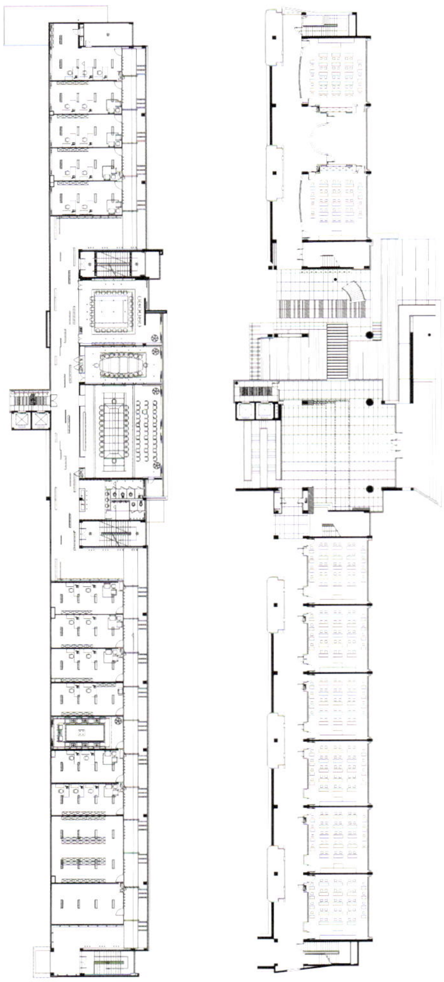

外立面以直线设计手法为基准，室内设计同样用此手法，点线面的运用做到室内外和谐统一，浑然一体。书本上条形码的重复，不规则的运用，无处不在地体现学校这一特定功能。

进入门厅，首先是做成台阶状的一层楼高的花坛，供学生休息时玩耍。顶上为石膏板与白色灯光片的结合，组成灯线。墙面为木化石干挂。原有立柱为方形，独立于大厅，参考整个空间后把其变形为一墙面，防火板包饰形成面，使空间更加统一。次厅同样采用直线顶，延续了大厅的设计风格。

原建筑设计为一长型走廊，北面为办公室。考虑到走廊太长会使人产生视觉疲劳，同时考虑到采光等因素，故平面做了大幅修改。中间会议室还是位于北侧，两侧办公室调整至南面，这样走廊长度大幅缩短，同时产生回廊，增加了空间感。为了更好地解决采光，办公室于走廊处采用了成片的玻璃隔断，踢脚线内缩，避免了后期积灰的可能性。

The exterior facade is designed in a technique of straight line while the indoor also applies the same technique. The application of dots, lines and surfaces has made up to the interior and exterior harmony and unification. The irregular repetitive application of the bar codes has shown everywhere the specific function of a school.

Once entering the hall, the parterre is designed into step style with one storey high for the students to have fun during the rest. The top is a combination of gypsum board and white light pieces to for into a light strip. The wall is hung with the wood fossil. The original vertical columns were square and are changed into a wall after reference to the whole space. The fire plates are packed into a surface. The annex hall has applied the straight-line ceiling and continued the design style of the hall.

The original building had a long corridor with the offices in the north. In consideration that people will have visual fatigue if the corridor is too long as well as the day lighting, the layout has been modified greatly. The middle conference rooms are still in the north while the offices on both sides are adjusted to the south so that the length of the corridor is greatly shortened. The cloister is also generated to increase the space senses. To better settle the day-lighting problem, the offices beside the corridor has applied the large pieces of glass as partition while the skidding boards are backed inwards to avoid the possibility of future dust accumulation.

线型构造媲美严谨的教学氛围。

The linear construction is comparable to the strict teaching atmosphere.

顶上为石膏板与白色灯光片的结合，组成灯线。

The ceiling is the combination of gypsum boards and white lighting flakes to form a lamp ribbon.

e-trading
Online Shopping
SUZHOU INDUSTRIAL PARK XINGWAN SCHOOL
Educational Vision
教育愿景
教育理想
办学特色
Teaching Characteristics
培养目标

大厅楼梯处。
At the staircase in the lobby.
光带在此成了最耀眼的主角。
The light ribbon becomes the most shining role.
旋转楼梯近景。
The close scenery of the circular staircase.
会议室一角。
One corner in the conference room.

先锋书店 Pioneer Bookstore

设计单位:南京筑内空间设计顾问有限公司 / 设计:陈卫新 / 面积:3000 m² / 主要材料:水泥地面、木质书架、乳胶漆、麻布 / 工程造价:150万元 / 坐落地点:南京 / 完工时间:2008年7月 / 摄影:张力

Design Unit:NANJING INTERIOR DESIGN CONSULTETION CO.,LTD. / Design:Chen Weixin / Area:3000 m² / Main Materials:cement floor、wooden bookshelf、emulsion paint、linen / Location:Nanjing / Project Cost:RMB 1 500 000 / Finish Time:Jul,2008 / Photography:Zhang li

先锋书店是中国著名的民营书店。五台山店是由原先的地下车库改造而成。考虑到投资的合理性与图书在书店空间中的重要作用，设计师几乎回避了所有贵重的材料。空间分割完全采用书架完成，重点保留了原先车库的地面及地面上的停车标线，使人流通道指示一目了然，同时也喻示了现代社会人与车一样，都需要一个“奔波”中的驻脚点。在入口的斜坡处，吊顶上方的著名人物的黑白摄影照片，仿佛是读者与他们的一次相互的心灵观望。序列状的立柱与灯为读者指明了方向，加强了仪式感。书永远是书店的主人，而书店是读者的精神道场。

It is a well-known individual bookstore in China. WuTai Shan Bookstore has been renovated and upgraded from the original underground garage. In view of reasonable in vestment and important effect of books in the spaces of a bookstore, the designer has almost evaded all valuable materials. All space was subdivided by shelves with the original floor of the garage and parking lines reserved, making traffic direction clear. At the same time it can drop a hint of similarity between men and cars and both need a parking point in the running. At the slope of the entrance, a photograph of a prominent public figure hung from the ceiling seems to be a view of soul by readers and it from each other. Sequence like columns and lamps can guide readers for ward to enhance a sense of ceremonies. Books are the owner of the bookstore for ever where readers can obtain nourishment for the mind.

书店门头。
Door header of the bookshop.
“十字架”只是隐喻书是人类的一种精神救赎。
The "Cross" is a metaphor that book is a spiritual salvation for the human beings.

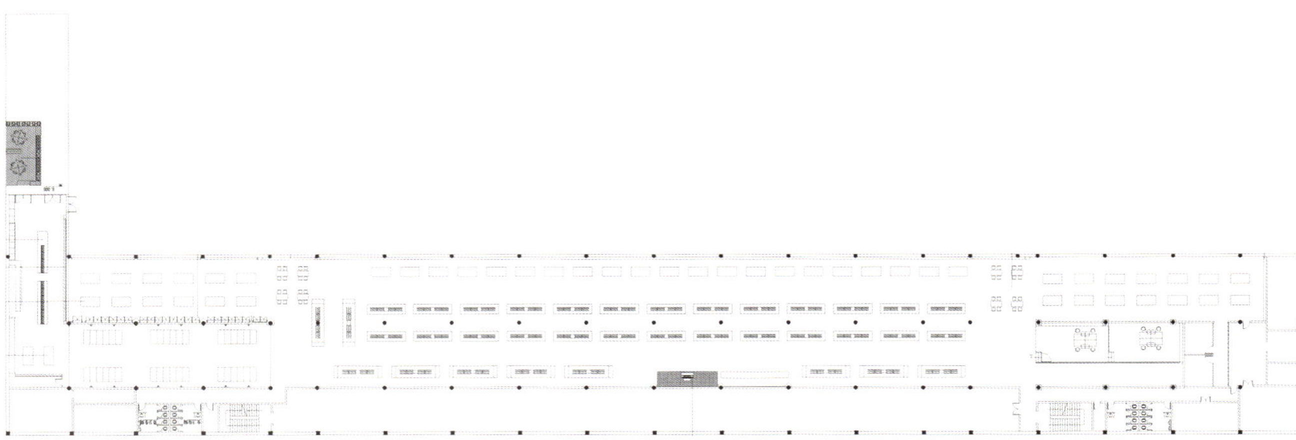

Xinjiekou
AVANT-GARDE
STRANGER ON THE EARTH
LIBRAIRIE
AVANT-GARDE
STRANGER ON THE EARTH

LIBRAIRIE
AVANT-GARDE
先鋒書店
大地上的异乡者

用“书”堆砌起来的服务台。
Service desk built with "books".
创意产品架。
Shelves for innovative products.
“读书”、“行路”，学者的两个方法。
"Reading" and "Touring" are two means of scholars.
序列状的立柱与灯具加强了入口过道的仪式感。
Sequential columns and lamps strengthen the ceremony feeling on the entrance corridor.

本色美术馆 True Color Museum

设计:陈瀚星 / 面积:12000 m^2 / 主要材料:钢结构、老青砖 / 坐落地点:苏州 / 工程造价:3500万元 / 完工时间:2008年10月 / 摄影:潘宇峰

Design:Chen Hanxing / Area:12000 m^2 / Main Materials:steel structure、old blue brick / Location:Suzhou / Project Cost:RMB 35 000 000 / Finish Time:Oct,2008 / Photography:Pan Yufeng

外立面玻璃和石材的搭配贴合本色二字。

The arrangement of the exterior facade glass and stone materials is integrated closely with the phrase "natural color".

会客区。

hospitality area.

本色美术馆位于苏州运河畔，依水而建，采用框架结构、局部钢结构相结合的方式，主体造型为三个立方体相连而成，东侧略带斜切面，方体之间形成南北及东西向景观及功能通道，局部挑空与延伸，产生戏剧性的空间效果。外立面用100多万块青砖凹凸拼贴而成，青灰色质感使整个建筑气势内敛而又饱含张力，与周边河流、湖泊、民居等自然生态相得益彰。

作为对外开放的公共艺术空间，本色美术馆在功能上有明确的区域划分，包括入口大厅、主题展区、办公区、艺术家工作室及公寓、会员俱乐部、咖啡厅、艺术品商店等。空间之间巧妙连接而相对独立，适合大展及主题个展。

环保概念与风水学在建筑理念中充分凸显，古木、老砖的大面积应用使整个建筑古朴自然，传统元素与当代观念完美结合。

Located at the riverside of the Suzhou Canal, True Color Museum is built by the combination of the frame structure and partial steel structure. The main part of the gallery is shaped like three connected cubes, and the east part is in the shape of a slightly diagonal plane. The three cubes form a landscape view in both vertical and horizontal directions. Partial single air and extension combine to create theatrical spatial results. The facade is covered by flashed bricks. The grey texture makes the building convergent and expansive, thus wonderfully complementing with the natural world such as rivers, lakes and houses.

As a public art location open for visitors, Originality Art Gallery is featured by different functions in different sections, including the entrance hall, theme exhibition halls, working area, member clubs, and artwork shops. Skillfully-connected locations are relatively independent, and are idea places for large-scale exhibitions and theme exhibitions.

Building concepts are fully focused on environmental protection and the art of geomantic omen. Ancient trees and old bricks are everywhere, which adds primitive simplicity to the entire art gallery. It is truly a perfect combination of traditional elements and modern concepts.

一扇扇木格窗户仿佛一个个古老故事的时光入口。

One after another wood pane windows look as if a time entrance to the old stories.

下沉地面上尽是斑驳的木格窗。
On the subsided ground is full of the blocky wood pane window.
热水瓶内胆也能成为不错的艺术品。
The internal vacuum flask chamber also can become a pretty good artwork.
灰色主打的厅内唯有灯光提醒人们回到现实。
Inside the hall where gray plays a leading role, only the light reminds the people to come back to the earth.
偌大的背景写真。
Such a big background picture.

北京韩美林艺术馆 Han Meilin's Art Museum

设计单位:杭州典尚建筑装饰设计有限公司 / 室内设计主持:陈耀光 / 面积:10000m² / 主要材料:石材、木材、玻璃、水泥板 / 坐落地点:北京 / 完工时间:2008年6月 / 摄影: 陈耀光

Design Unit:Hangzhou Dianshang Building Decoration Design Co., Ltd / Principal Interior Design:Chen Yaoguang / Area:10 000m² / Main Materials:stone、wood、glass、cement board / Location:Beijing / Finish Time:Jun,2008 / Photography:Chen Yaoguang

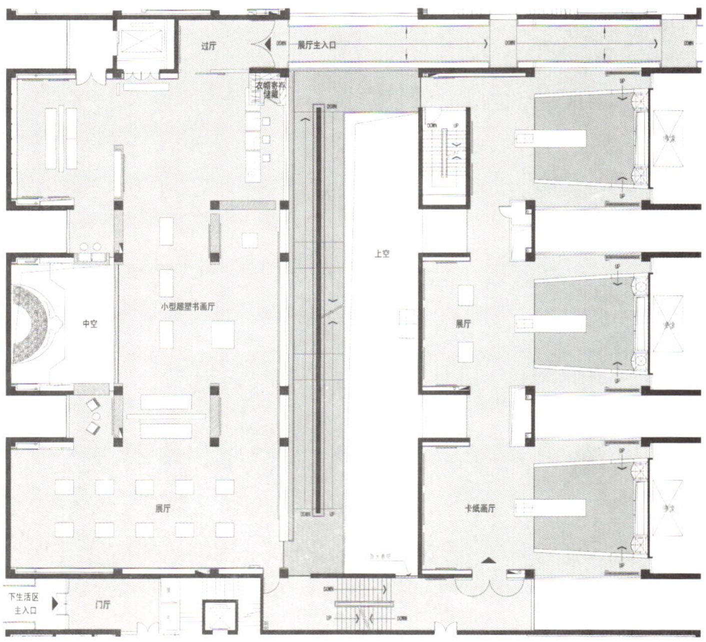

本项目建筑由中国建筑设计研究院崔恺大师主持设计，我们作为室内设计师，与建筑设计师共同合作完成这一作品。建筑主体周边设置了下沉式水院，核心的展览区位于整个建筑的中部。展厅设计成4m和6m两种高低不同的空间。其中，建筑最中心区有一个近12m的采光高厅。除了展示厅、雕塑厅、国画书法厅、陶瓷厅、卡纸画厅、工艺美术展厅以外，主入口处还设有演示交流厅和纪念品商店。采用了“宁静、悠远、宽广、大气、东方、民族”的设计定位和设计理念。作品的展示空间应该是平静、从容和内敛的。在空间设计元素上，应控制起伏、把握平静，材料、色彩、肌理及形态次于展品。艺术馆、博物馆，所有的展示场馆空间材料最理想的表现形式应是让所有的材料不太具有表现性。它的意义、它的存在，都应该是为了突出空间的主题——展品所服务的。另外，材料的质感、明度、它的肌理气质应该和这个建筑浑厚坚实的外观浑然一体。

The architecture of this project is designed under the leadership of the master Kai Cui from China Architecture Design&Research Group while as the interior designers, we have coorperated with him to accomplish this work. The subsidence-type water courtyard is arranged around the main building with the core exhibition area is located in the middle of the whole building. The exhibition hall is designed into two kinds of spaces in different heights – 4m and 6m high, among which there is an approximately 12m high day-lighting hall in the most central area of the building. Apart from the exhibition hall, sculpture hall, the hall for traditional Chinese paintings and calligraphies, ceramic hall, the hall for plane paintings and the exhibition hall for industrial artworks, there are also the demonstration and communication hall as well as the remembrance shop. The design is orientated at "serenity, remoteness, wideness, grandness, east and nationality". With regard to the design elements for space, the waves shall be controlled and the calmness shall be grasped. The materials, colors, textures and forms are secondary to the exhibits. The most ideal expressive form of the space materials in the exhibition hall is to make all materials with little expressiveness. Its existence shall be purposed to stress the theme of the space – serve for the exhibits. Furthermore, the qualities, brightness and textures of the materials shall be perfectly and naturally integrated with the simple and vigorous as well as firm appearance of this building.

室外下沉庭院1。
The outdoor subsided courtyard 1.
室外下沉庭院2。
The outdoor subsided courtyard 1.

雕塑厅局部。
Part of Sculpture Hall.
雕塑厅大佛。
Buddha in Sculpture Hall.
雕塑厅全景。
Full View of Sculpture Hall.

中国人民解放军第98医院 Chinese PLA No.98 Hospital

设计单位:南京亚明环境艺术设计有限公司 / 设计:孙亚明 / 面积:30500 m^2 / 主要材料:镜面不锈钢、玉晶石、美国灰麻、瓦楞板、白色地转 / 坐落地点:湖州 / 工程造价:4200万元 / 完工时间:2008年5月

Design Unit:Nanjing Yaming Environment Art Design Corporation / Design:Sun Yaming / Area:30500 m^2 / Main Materials:mirror stainless steel、spar、gray hemp、corrugated sheet、white ground loop / Location:Huzhou / Project Cost:PMB 42 000 000 / Finish Time:May,2008

整个设计色彩淡雅，以明快、亲切、柔和的自然色，替代白色所带来的冷酷、肃穆的医疗形象，增加医疗空间的生机与活力，也为患者提供了恬静、舒适的救治环境。通过简洁、现代的设计手法，塑造具有特色、富有情趣、体现人文关怀、充满文化内涵的医疗室内空间。

The colors of the whole design are quietly elegant. The bright, friendly and soft colors are used to replace the cool and solemn medical image brought by white and increase the vitality and vigor of the medical space as well as provide a quiet and comfortable curing environment to the patients. It has created a unique and vivid interior medical space with reflection of human cultural considerations and full of cultural meanings through the concise and modern design technique.

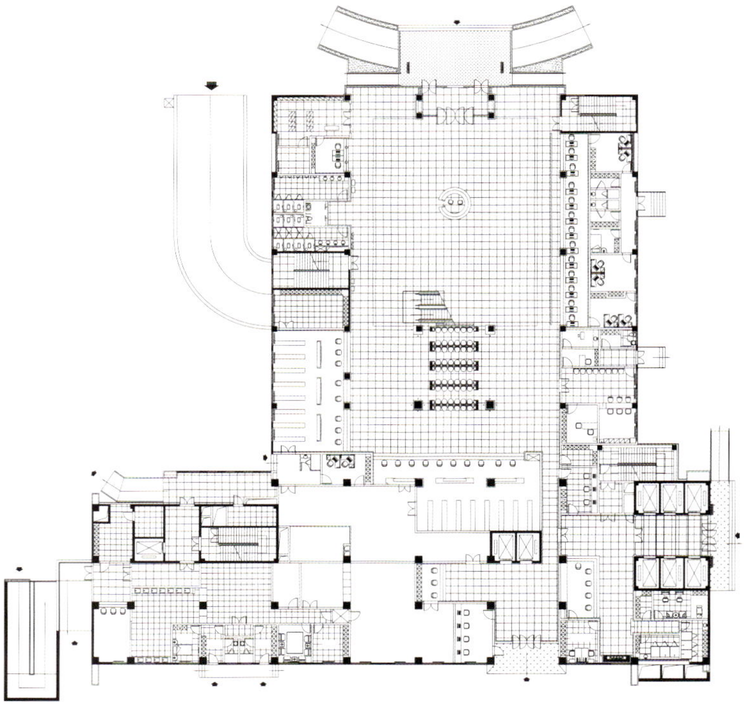

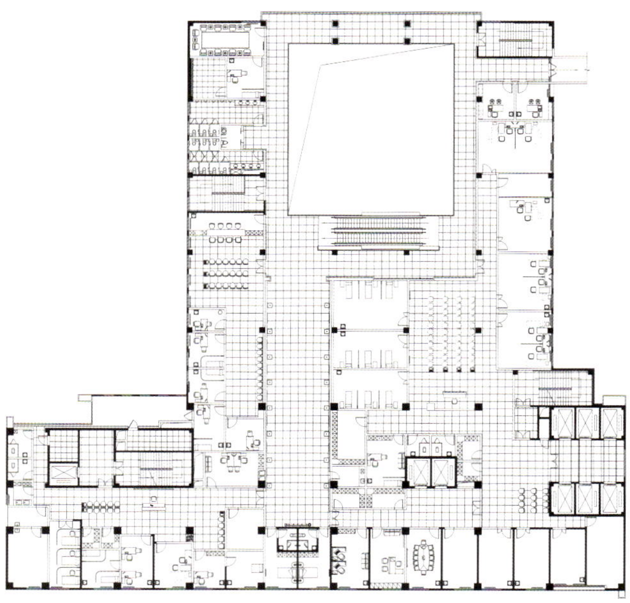

通透统一的中庭。
The transparent and unified atrium.
蓝色背景让人心情平静、宽广。
The blue background makes people calm and broad.
秩序井然的设计手法。
The well-ordered design technique.

JOHNSON
CONTROLS

提示
请保管好自己的财物
护士站
15:40
内 科
17–19

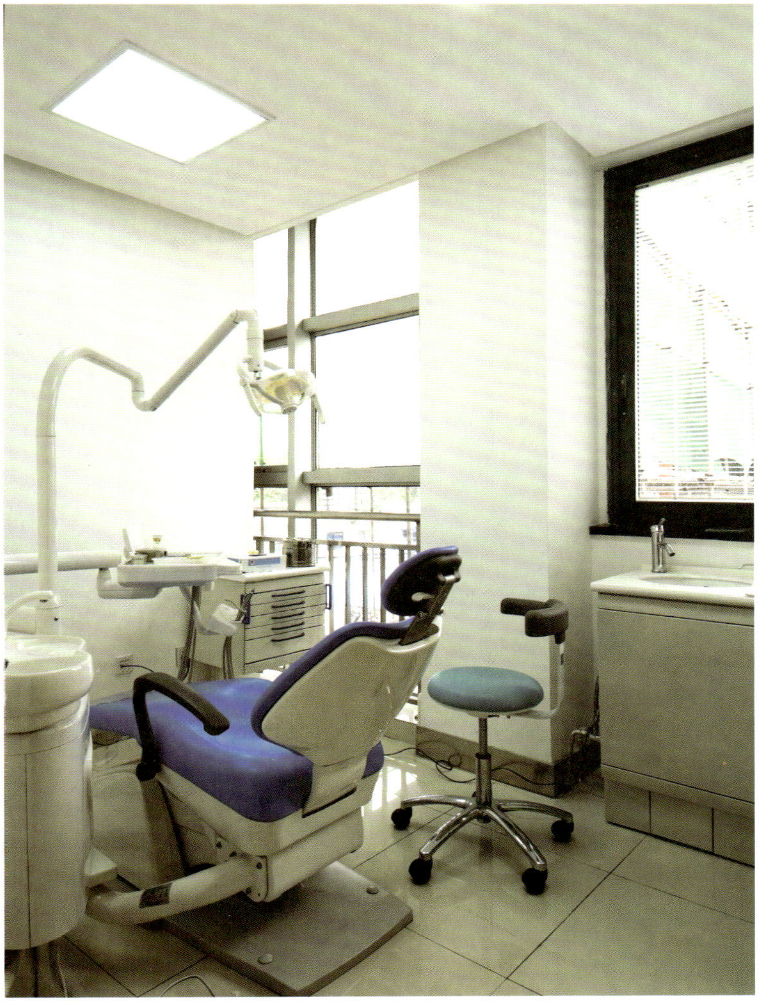

过道连接电梯处。

The corridor connected with the elevator.

护士站服务台采用有色磨砂玻璃，清新、自然。

The service desk at the nurse station has applied the colored frosted glass, fresh and natural.

灯光的设计完全考虑其实际功能性。

The design of the light has taken full consideration of its practical functionality.

落地玻璃的病房消除人们的紧张感。

The sick ward with floor glass eliminates the people' s nervousness.

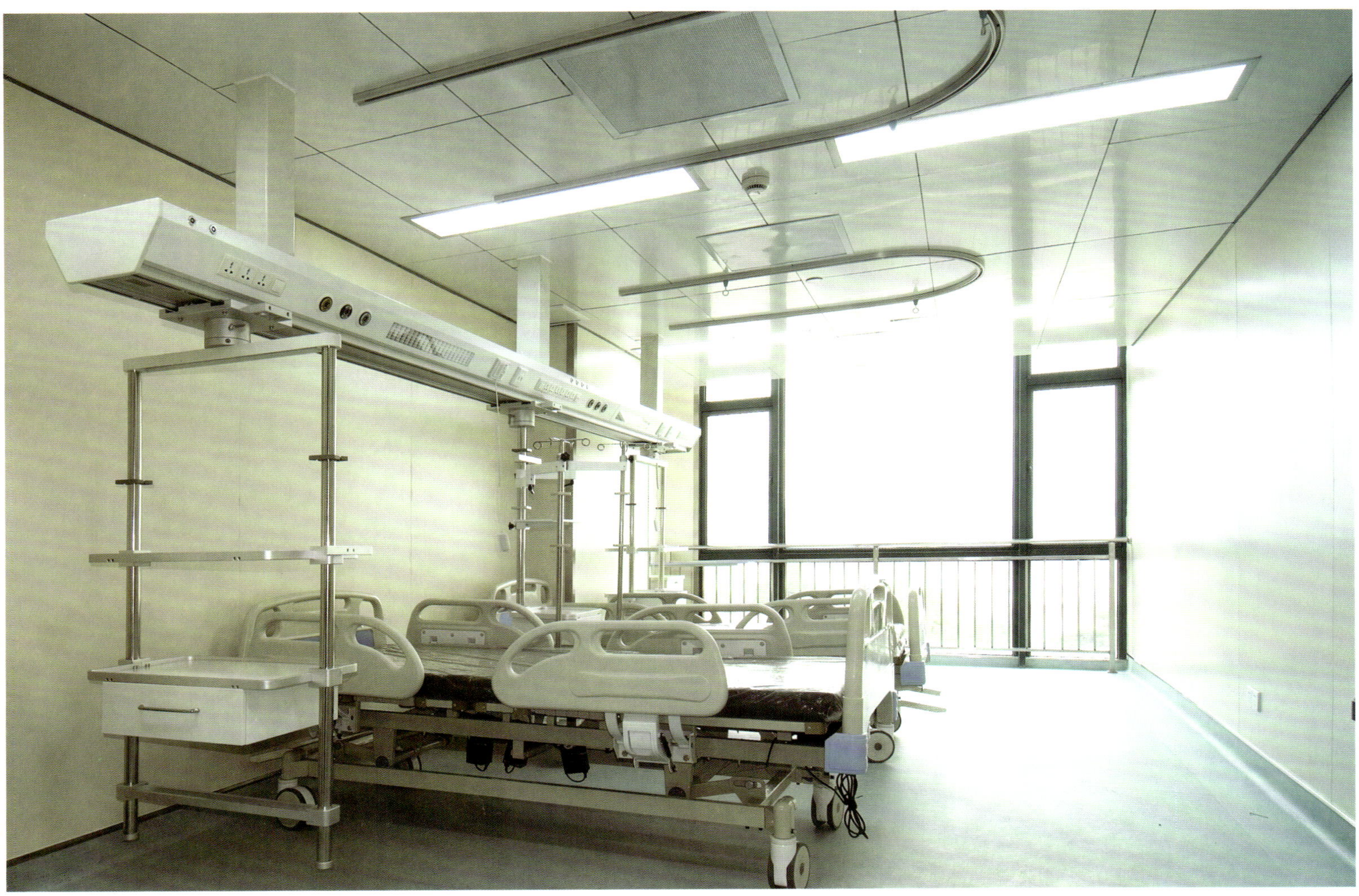

曹 成

深圳深港建设及深圳汇博设计总监、高级室内设计师、高级工程师、国家一级专案经理、环境艺术设计师。

陈德坚

毕业于英国李斯特迪蒙福大学，主修室内设计。曾于英国曼彻斯特城担任室内设计师，后成立德坚设计。香港室内设计协会（HKIDA）主席，香港设计中心董事及香港贸易发展局顾问。

陈广暄

深圳设计师，1995年受聘于深圳市装饰工程工业总公司江苏分公司，担任设计部负责人。1998年于南京组建斑马装饰设计工作室。2003年创办广暄工程设计顾问有限公司。2006年与业界伙伴组建多元设计联合体"优联设计机构"，业务涉及商业业态策划，工程技术顾问，VI设计及推广。

陈 骏

毕业于汕头大学美术设计系本科。蓝鲸室内设计有限公司经理、创意总监。

陈品豪

毕业于清华大学美术学院，宁波大学艺术与传媒学院、宁波城市学院副教授。宁波柏天设计公司设计总监，中国建筑学会室内设计分会会员，ICIAD（国际室内建筑师与设计师理事会）宁波地区理事。

陈石林

1997年毕业于中国美术学院环境艺术系，2000年合伙创立杭州正午装饰设计公司，2005年受巴黎国际艺术城邀请，赴法国游学。从事高档娱乐、餐饮、会所、酒店、宾馆等大型公共场所的设计。主要作品有美国纽约LINDEN PLACE娱乐会所、美国纽约SHANG HAI LEE餐厅、上海顺风大酒店、上海金碧辉煌娱乐城、杭州银乐迪KTV、杭州世贸西湖四季酒店、萧山凯撒皇宫娱乐城。

陈耀光

杭州典尚建筑装饰设计有限公司创意总监、总经理。中国建筑学会室内设计分会、中国建筑装饰协会设计委员会、中国陈设艺术专业委员会，副会长。美国Hall of Fame名人堂2007中国（含港澳台）首批7位成员之一。浙江省建筑装饰行业协会设计委员会，会长。1996年-2005年，连续十届中国室内设计大赛七次荣获一等奖。2007年度中国十大样板房设计师。

陈旭东

毕业于吉林艺术学院美术系，获CIID2004年-2006年度全国百名优秀室内建筑师荣誉称号，吉林艺术学院设计学院教授、硕士生导师。

陈 武

新冶（香港）设计工程有限公司董事，深圳市新冶组设计顾问有限公司创办人，中国区"夜时尚"大型跨界联盟活动发起人，中国民族文化促进会会员，北京欢乐时空动漫学院客座教授，国际室内建筑师&设计师理事会理事，中国室内装饰协会会员，IFI会员，当代中国设计与价值合作事业联盟(UCDVR) 专业设计顾问。

端木芸萱

上海基美设计有限公司艺术总监，新西兰基督城理工大学室内设计本科，澳洲新南威尔士大学室内建筑本科，英国布莱顿大学室内设计硕士，新西兰籍。

范 江

高得设计公司设计总监兼负责人。作品常见于《室内设计与装修》、《现代装饰》、《中国建筑装饰装修》、《美国室内》中文版等专业杂志。

崔华峰

崔华峰空间设计艺术顾问工作室主持人，研究方向：当代·东方·印象。

符军

RYB·三原色建筑装饰设计院院长兼首席设计师。高级环境艺术设计师，高级室内建筑师，吉林大学珠海学院艺术设计系兼职教授，珠海城市职业技术学院客座教授，全国百名优秀室内建筑师，首届广东省装饰行业功勋设计师，中国建筑学会室内设计分会理事兼三十（珠海）专业委员会理事长，广东省环境艺术设计协会副会长，珠海市室内设计协会会长，广东省装饰设计专家委员。

方振华

室内建筑设计师，方振华设计（香港、上海）有限公司董事，香港室内设计协会会长(1993-1995)，香港设计师协会执委(1997-2006)，中国陈设艺术专业委员会副主任、酒店协会常务理事、美术学院特聘教授、苏州大学特聘教授，美国哥伦布艺术设计学院学士、国际室内设计学会注册专业会员，英国特许设计师学会注册专业会员，国际室内建筑设计师联盟会员，美国建筑师学会准会员。

冯羽

毕业于天津美术学院环境艺术设计本科，深圳市大羽环境艺术设计有限公司设计总监。

方钦正

建筑师，台湾人。毕业于英格兰的一所学校，在中国有7年工作经验。负责亚洲地区酒店、零售及活动等有关的各种项目。在施工方面具有丰富的经验。

高龙

第四届上海市十大优秀青年室内设计师；上海理工大学艺术设计学院环境艺术设计专业教师；上海熠彩装饰设计有限公司总设计师；上海汉星艺术设计有限公司总设计师；CIID会员；中国室内装饰学会陈设艺术专业委员会会员。

高文安

高文安设计有限公司董事长。

高轶

1999年取得平面设计专业学士学位，1999年-2003年从事平面设计工作期间，专业设计企业标志和品牌视觉形象，2004年取得室内设计专业学士学位，2004年创办南京目达室内设计有限公司，任设计总监。CIID会员。2009年创办南京高轶装饰设计有限公司。

耿治国

出生于台湾，东海大学建筑系毕业，1998年创立台北飞形设计，2001年创立上海飞形设计。

谷腾

中国室内高级设计师。IFI国际室内设计师/建筑师联盟会员，中国室内设计学会会员。荣获第六届中国室内设计双年展室内设计大赛银奖，14届香港亚太设计大奖办公类金奖，2007《现代装饰》国际传媒奖室内设计大奖，2007中国室内设计大赛最具创意奖等。

张恒祥

苏州金螳螂建筑装饰股份有限公司第五设计院所长。
主要作品：国家机关干部培训中心、杭州中国丝绸博物馆、常熟时风国际大酒店、苏州万科玲珑湾样板房、苏州金澄明珠样板房、北京首都博物馆（新馆）民俗展馆、吴江同里湖度假村、上海汽车博物馆、常州博物馆、无锡江苏省太湖干部疗养院等。

韩冠恒

2001年毕业于黑龙江省建筑职业技术学院建筑与城规系。现任哈尔滨唯美源装饰设计有限公司经营部经理。中国建筑学会室内设计分会会员。室内设计师。

洪忠轩

HHD假日东方国际酒店设计机构首席设计师，国际注册设计师，香港室内设计协会深圳代表处理事委员、ICAD国际设计师。被世界酒店领袖中国会评为“世界酒店·2008中国杰出酒店设计师”；2003年荣获中国“CIID最佳室内设计师奖”；中国国际饭店业博览会“最佳饭店设计师”。

霍承显

注册高级室内设计师，高级室内建筑师，珠海空间印象首席设计师，中国建筑学会室内设计分会会员，广东省“岭南杯”百名优秀设计师，2008 CIDA“精英奖”设计师。

姜 峰

建筑学硕士、国务院特殊津贴专家、教授级高级建筑师、高级室内建筑师。姜峰室内设计有限公司总经理兼总设计师、中国建筑学会室内设计分会副会长、中国建筑装饰协会设计委副主任。先后荣获中国室内设计师年度十大人物、深圳市十大杰出青年、中国室内设计20年设计功勋奖、中国国际设计艺术观摩展终身艺术设计成就奖等荣誉。

姜湘岳、 吴海燕

毕业于南京艺术学院工艺系装饰专业，同年任职于江苏省建筑科学研究院，1997年任江苏省建科院装饰设计所所长，2001年至今任江苏省海岳建筑装饰工程设计有限公司总设计师和南京金陵国际装饰设计工程实业有限公司设计总监。致力于研究酒店的设计及发展。CIID会员，高级室内建筑师。

靳全勇

1996年毕业于哈尔滨大学环艺系。室内建筑师。现任哈尔滨唯美源装饰设计有限公司设计师。

张 晔

本科毕业于重庆建筑大学建筑学室内设计专业，清华大学建筑学院工程硕士。任中国建筑设计研究院环境艺术设计研究院副总建筑师，教授级高级建筑师。曾获中国建筑装饰协会评选的“杰出青年室内建筑师”、“全国百名优秀室内设计师”、“最具影响力室内设计师”称号，2004年“美国室内中文版”杂志十大封面人物。第三届中国国际设计艺术观摩展十大最具观摩展示奖。

琚 宾

就读于中央美术学院设计学院，主修室内设计，毕业后在香港、深圳、北京从事专业室内设计，主持方案设计、主题性酒店设计专家，致力于研究中国文化在建筑空间里的运用和创新。并与国内国际多位知名建筑师合作设计项目。以个性化，独特的视觉语言来表达设计理念。以全新的视觉传达来解读中国文化元素。

李浩澜

毕业于鲁迅美术学院环境艺术系。室内建筑师。中国建筑学会室内设计分会会员。

李益中

毕业于大连理工大学建筑学专业，1998年创立深圳市派尚环境艺术设计有限公司，2007年出版著作《样板生活》，现为中国建筑学会室内设计分会理事及第三专业委员会副会长。近年来主要致力于地产商的样板房、售楼处、会所等为主的专业室内设计，并在此领域积累了大量的成功案例和经典作品。

梁宇曦

毕业于佛山大学艺术系室内设计专业，高级室内建筑师，中国建筑学会室内设计分会理事，CASCANDO(奇思度)设计策划设计总监。

梁志天

1997年创办梁志天设计师有限公司，并于上海、北京、广州及深圳设立办公室。2001年起发展家具及产品设计，作品包括《一摯》、《方元》、《玲珑》及《天地》系列。七度获得素有室内设计奥斯卡之称的Andrew Martin International Awards殿选为全球著名室内设计师之一；其作品亦连续数年于“亚太室内设计大奖”夺得多项最高殊荣，并获得多个国际酒店设计奖项,如美国的Gold Key Awards，及Hospitality Design Awards，曾获超过70项国际和亚太区个人及设计奖项。

林开新

福州林开新室内设计有限公司负责人，CIID第八委员会理事，全国百名优秀室内建筑师。

凌子达

毕业于台湾逢甲大学建筑系。先后在台中及台北等城市实践自己对建筑和室内设计的理念。2001年到上海发展，并成立了达观建筑室内设计事务所。2005年荣获上海市第一届“十大优秀青年室内设计师”。于北京荣获“2005中国设计业青年百人榜”。个人作品入选《2004亚太室内设计大奖作品选》、《2005中国室内设计年鉴》、《2005国际室内设计师上海别墅专辑》。

刘 志

法国国立工艺学院CNAM-IIM项目管理硕士。现任唯美源设计北京分部设计总监兼经理，CIID会员。CIID全国百名优秀室内建筑师。2009-2010年中国设计博览会专家委员。

卢 迅

东莞形上几何室内设计有限公司总设计师，中国建筑学会室内设计分会理事，广东省环境艺术设计行业协会专家组专家。2007年荣获“2004-2006年度全国百名优秀室内建筑师”称号。

吕邵苍

中国百名优秀室内建筑师。金螳螂设计八分院院长 。中国建筑学会室内设计分会理事。中国建筑学会室内设计分会第36（无锡）专委会副主任。国际ICIAD室内设计学会理事。国际IFI设计学会会员。

吕永中

毕业于上海同济大学建筑系，现任教于同济大学建筑与城市规划学院艺术设计系，上海市工业设计协会家具设计专业委员会副主任，中法合资唯品设计董事长及设计总监，半木原创家居品牌创始人及主持设计师。长期从事空间设计及家具研究，挖掘中国传统文化内涵，在现代设计的家具中注入“平和、内敛”的东方性格，以回归宁静的本真状态及听慧的生活态度表达其哲学意境。

骆志兴

广州新锐设计师，广州力思园林与室内工程设计有限公司执行董事、首席设计师。

裴晓军

毕业于天津工艺美术学院环境艺术系，美国PDA室内设计事务所负责人，高级室内建筑师、高级室内设计师，IFDA国际室内装饰设计协会国际资深会员，中国住宅委专家库专家，中国建筑学会室内分会理事，江苏省勘协建筑装饰及环境艺术设计专业委员会秘书长，南京市室内设计学会副会长，南京工业大学艺术设计学院兼职教授，中国美术家协会江苏分会会员。

迫 庆一郎

1996年东京工业大学研究生毕业，1996年-2004年就职于山本理显设计工场，2004年成立SAKO建筑设计工社（中国北京），2004年-2005年赴哥伦比亚大学担任客座研究员，日本文化厅外派艺术家驻外研修员。

秦岳明

深圳市朗联设计顾问有限公司设计总监，1990年建筑学专业毕业，1994年起开始涉足室内设计。1999年创建朗联设计公司，多年来带领朗联团队致力于酒店会所、办公空间及楼盘示范单位的设计与研究，成功完成多个商业项目并在国内及亚太获得众多设计奖项。

宋国梁

中国百名优秀室内建筑师，ICIAD国际室内建筑师与设计师理事会宁波区理事，IFI国际室内建筑师设计师联盟会员，中国建筑学会室内设计分会会员，HBS+宁波红宝石装饰设计有限公司总设计师，2004年获全国百名优秀室内建筑师荣誉称号。

宋微建

上海微建（Vjian）建筑空间设计有限公司董事长兼首席设计师，中国建筑学会室内设计分会理事，中国建筑学会室内设计分会专家委员会委员，中国建筑学会室内设计分会第十二专业委员会会长，苏州科技大学建筑学院客座教授，全国杰出中青年室内建筑师，全国有成就资深室内建筑师，中国十佳酒店设计师，2006中国室内设计师年度封面人物，2008国际传媒杰出设计师。

孙 亮

毕业于成都市教育学院，现任“成都优佰文化传播有限公司”设计总监、“真业集团（香港）有限公司”设计总监。

孙亚明

南京亚明环境艺术设计有限公司设计总监、总经理。主设项目：南京夜上海酒店、南京军区南京总医院、南京好记大酒店、南京永卿食府酒店、中国人民解放军第98医院、广州军区武汉总医院。

陆文星

2004年创立陆文星室内设计事务所至今，作品覆盖酒店、会所、办公、样板房、别墅、商业等多个领域，尤以设计各类品牌个性化商业项目见长。已有众多作品刊登于《亚太室内设计大奖》、《中国室内设计年鉴》等专业刊物。

童武民

高级室内建筑师，全国百名优秀室内建筑师，江西省著名室内设计师，江西首位获得大型房地产（千户建筑群）样板房设计资格设计师，IFI会员，中国建筑学会室内设计分会会员，中国建筑学会室内设计分会第27（江西）专业委员会副秘书长、专家评委、委员。

万宏伟

毕业于四川美术学院。全国百名优秀设计师。宁波市汉文设计工作室总经理、设计总监，CIID宁波专业委员会理事，第三届中国国际设计艺术观摩展重要奖项—2007室内设计十大新锐人物。

王 宏

BDM 公司首席设计师，1991年毕业于中国重庆大学建筑学院室内设计专业；英国伦敦艺术大学切尔西艺术学院环境艺术设计专业 硕士学位。2007年其参赛作品“运河岸上的院子”成为中国大陆第一位获得英国著名室内设计奖 —“Andrew Martin”。

王严民

黑龙江省佳木斯市豪思环境艺术顾问设计公司首席设计师。大连市豪思环境艺术装饰设计机构主案设计师。国家注册高级室内设计师、室内建筑师。中国室内装饰协会专业会员。中国建筑学会室内设计分会会员。2004年-2006年度全国百名优秀室内建筑师。黑龙江省艺术设计协会会员。

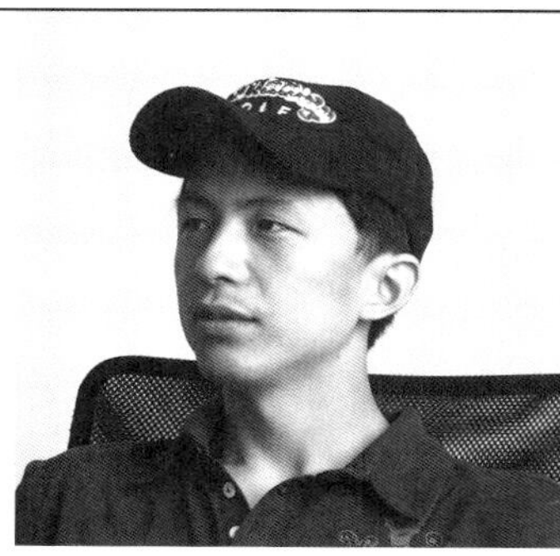

王严平

任职于苏州国贸嘉和建筑工程有限公司。工程师、室内建筑师。

王祎华

苏州金螳螂建筑装饰股份有限公司第一设计分院一所负责人；CIID、中国建筑装饰协会及中国饭店协会中国十佳酒店设计师；室内建筑师、工艺美术师；中国建筑学会室内设计分会会员；高星级酒店、商业地产设计策划研究人。

王严钧

黑龙江省佳木斯市豪思环境艺术顾问设计公司设计总监。国家注册高级室内设计师、室内建筑师、中国室内装饰协会专业会员、中国建筑学会室内设计分会会员、IFI（国际室内建筑师/室内设计师联盟）会员、黑龙江省室内设计专业委员会理事、黑龙江省艺术设计协会理事、佳木斯市艺术设计协会副会长。

吴 钢

WSP主设计师、董事总经理。同济大学风景设计专业本科，德国卡尔斯鲁厄大学建筑学硕士、博士；1994 年为慕尼黑西门子房产管理公司建筑设计部设计主持人、亚洲项目主管。1997 年作为合伙人成立慕尼黑 WSP 建筑师事务所，1998 年回到中国北京成立 WSP 建筑师北京事务所。

吴君华

中国建筑学会室内设计分会会员，国际室内建筑师与设计师理事会会员，现任宁波翰禹·印加空间设计机构执行懂事兼设计总监。

杨邦胜

米兰理工大学设计学院硕士，YAC杨邦胜酒店设计顾问公司董事长、设计总监，亚太区著名酒店设计师，中国设计业十大杰出青年，中国建筑学会室内设计分会常务理事。从事酒店室内设计工作15年来，兼收并蓄，致力于文化个性酒店的设计与探索，曾主持众多国际知名设计项目，作品涵盖凯宾斯基、洲际、喜达屋、豪生等国际品牌，并屡获国际国内大奖。

曾建龙

毕业于福州大学工艺美术学院和美国国际联盟大学艺术设计学院、中国建筑学会室内设计分会会员、室内建筑师、ICIAD国际室内建筑师与设计师理事会理事。荣获奖项有2005年海峡两岸三地别墅空间设计三等奖、2007年桑格拉杯首届中国建筑学会室内设计分会大奖赛工程类一等奖及方案类二等奖、金羊奖2007年度中国十大设计师奖等。

张 瑛

WSP主设计师、董事，同济大学建筑城规学院风景设计专业本科;德国卡尔斯鲁厄大学建筑设计硕士。

陈 凌

WSP董事、主设计师，毕业于武汉工业大学，后赴巴黎第一建筑学院（原巴黎美术学院）学习，期间在多家著名建筑师事务所从事建筑、城市规划及室内外环境设计工作。2005年被评为中国十大新锐建筑师。

张奇永

1999年就读于黑龙江建筑职业技术学院环艺系。2007年就读于哈尔滨工业大学艺术设计系。现任哈尔滨唯美源装饰设计有限公司设计师兼建筑摄影师。CIID会员。室内设计师。

张 灿

高级室内建筑师。全国杰出中青年室内建筑师。全国百名优秀室内建筑师。IAI亚太建筑师与室内设计师联盟理事。中国建筑学会室内设计分会理事。中国建筑学会室内设计分会第四专业委员会委员。成都市建筑装饰协会副会长。四川创视达建筑装饰设计有限公司 负责人。

张慧勇

毕业于福建师范大学。福州品川设计顾问公司联席设计总监。ICIAD国际室内建筑师与设计师理事会华南区及福建代表处理事长。福建省住宅产业商会理事。中国室内建筑师。中国室内住宅设计师。中国建筑学会室内设计分会会员。2005年度福建省优秀室内建筑设计师。2004年度福州市十佳家装设计师。

周浩明

高级工程师，苏州国贸建筑装饰工程有限公司总设计师，国际注册高级室内设计师，全国杰出中青年室内建筑师。

周 际

全国资深室内建筑师。中国建筑学会室内设计分会会员。中国建筑学会室内设计分会第三专业委员会委员。深圳市装饰行业专家库成员。深圳市室内设计师协会常务理事。深圳市图人设计有限公司董事长、首席室内设计师。2004年被评为“全国百名优秀室内建筑师”和“全国杰出中青年室内建筑师”。

周 捷

高级设计师。1998年毕业于中国科技经营管理大学，室内设计专业。
设计作品：IFC、德勤会计师事务所，德国电讯，万达广场，西门子西安，ABB，赛贝斯北京，深圳发展银行，NCR，东亚银行，中信银行等。

吴 纯

毕业于苏州工艺美术学校室内设计专业。工程师。苏州美瑞德建筑装饰有限公司副总设计师。

李勇强

毕业于苏州工艺美术职业学院（原苏州工艺美术学校）装潢设计专业。苏州金螳螂企业美瑞德建筑装饰有限公司设计一所所长。2006年度全国杰出中青年室内建筑师。2005年度中国优秀酒店设计师。2005年IAID最具影响力的中青年设计师。

陆 逊

苏州苏明装饰有限公司常务副总兼设计分公司经理。工程师。2007年荣获全国建筑工程装饰奖设计师荣誉；2007年当选为江苏省装饰装修行业协会设计委员会副主任委员；2007年荣获“江苏省十佳青年室内设计师”荣誉称号；2006年荣获“全国杰出中青年室内建筑师”荣誉称号；2004年荣获中国建筑装饰协会“高级室内建筑师”资格证书。

郭 继

毕业于福建师范大学室内设计专业，福州品川设计顾问公司联席设计总监，IFI国际室内建筑师/设计师联盟会员，中国室内建筑师，中国室内住宅设计师，中国建筑学会室内设计分会会员，2005年度福建省优秀室内建筑设计师，2004年度福州市十佳家装设计师，2006年度 全国百名优秀室内建筑师。

张丰义

杭州金白水清悦酒店设计公司董事长、设计总监。

王仲文

1997年毕业于哈尔滨市轻工学院室内设计系。室内建筑师、CIID会员、哈尔滨市唯美源装饰设计有限公司创意室主任兼设计师。

盖晓军

毕业于苏州工艺美术职业技术学院室内设计系。苏州金螳螂建筑装饰有限公司陈卫潭工作室一所所长。
主要作品：苏州图书馆、常州大酒店、无锡会展中心、苏州工业园区地税办公楼、江苏省省委办公楼、溧阳天目湖宾馆、南京奥体中心体育场、常州香格里拉大酒店、江苏大厦新楼、乌镇度假酒店、苏州出入境检验检疫局办公楼、北京远通维景国际大酒店等。

刘 桦

毕业于无锡轻工业学院室内设计系。苏州金螳螂建筑装饰有限公司陈卫潭工作室三所所长。主要作品：成都飞机制造研究中心办公楼、苏州新区科技大厦、苏州新城花园酒店、常州金陵太湖湾酒店、常州金鼎国际大酒店等。

于强室内设计师事务所主创设计师团队合影。

从左至右（双排是先前而后）：杨 冰 林敏玲 王冠 毛 桦 于 强 詹富乔

张 楠

1994年起任南京佳的设计公司总设计师，2004年至今任南京远境空间环境设计公司总设计师。

王建强

杭州金白水清悦设计公司总经理、设计一所所长。

唐 凡

高级室内建筑师。大木设计中国发展有限公司（北京公司）负责人、设计总监、CIID会员。IFI国际室内建筑师联盟会员。1993年毕业于哈尔滨师范大学美术系。2007年毕业于清华大学建筑工程与设计高研班。2009年法国国立工艺与技术学院设计项目管理硕士。

蔡 鑫

HKG总经理助理、主创设计师。哈尔滨工业大学本科。灵山梵宫项目主创设计师。

鱼晓亮

HKG设计总监助理、主创设计师。西安建筑科技大学本科。灵山梵宫项目主创设计师。

翁芝青

HKG主创设计师。中国美术学院本科。灵山梵宫项目主创设计师。

卢志刚

上海米丈建筑设计有限公司总经理，总建筑师、高级工程师、国家一级注册建筑师。毕业于重庆建筑大学。主要代表作有：上海高级人民法院审判法庭办公楼、杭州天工艺苑、贝安公司总部办公楼、上海中融国际商厦、宁波物流大厦等。著有《城市取样1×1》及《米丈建筑地图》系列丛书。获首届上海市青年建筑师新秀奖金奖，第五届中国建筑学会青年建筑师奖。

郭轲蔚

高级室内建筑师。金螳螂浙江设计研究院院长。郭轲蔚酒店设计事务所。金螳螂建筑装饰股份有限公司副总设计师。浙江省建筑装饰协会设计委员会副会长。

张成喆

1997年旅居巴黎，1999年定居上海，成立IADC国际涞澳设计公司，2007年出版个人作品集《喆思空间》。多次荣获国内外重要设计奖项：2006年度亚太精英设计师奖、2008年度设计师网络传媒奖、2008年度金外滩最佳居住空间奖等。

主编
陈卫新

编委（排名不分先后）
陈耀光、陈南、高蓓、黄玉枝、张乘风、孙天文
沈雷、宋桥林、王琼、王兆明、吴海燕、叶铮

国书在版编目（CIP）数据

2009中国室内设计年鉴 /《2009中国室内设计年鉴》编辑组编. —沈阳：辽宁科学技术出版社，2009.11
ISBN 978-7-5381-6142-7

Ⅰ.2… Ⅱ.2… Ⅲ.室内设计—中国—2009—年鉴
Ⅳ.TU238-54

中国版本图书馆CIP数据核字（2009）第179322号

出版发行：辽宁科学技术出版社
（沈阳市和平区十一纬路29号 邮编：110003）
印 刷 者：上海美雅延中印刷有限公司
经 销 者：各地新华书店
幅面尺寸：230mm × 300mm
印　　张：88
插　　页：8
字　　数：300千字
印　　数：1~1200
出版时间：2009年 11 月第 1 版
印刷时间：2010年 2 月第 1 次印刷
责任编辑：陈慈良
封面设计：赵宝伟
版式设计：赵宝伟
责任校对：周　文

书　　号：ISBN 978-7-5381-6142-7
定　　价：560.00元 (1、2) 册

联系电话：024-23284360
邮购热线：024-23284502
E-mail:lnkjc@126.com
http://www.lnkj.com.cn
本书网址：www.lnkj.cn/uri.sh/6142
国内直销总代理：曾江福　13564489269